高职高专"十四五"规划教材

智能网联汽车 ROS 实战入门

曾子铭
深圳市元创兴科技有限公司 编著

北京航空航天大学出版社

内容简介

本书以智能网联汽车 ROS 基础知识为核心内容,介绍智能网联汽车开发平台 ROS 的基本概念和使用方法,包括节点、ROS 通信机制、坐标转换等内容。通过学习 Linux 操作指令、ROS 指令、ROS 通信管理机制、可视化工具和智能网联汽车组成架构等知识,完成智能网联汽车虚拟化模型的搭建。

本书适合高职院校智能交通技术专业、汽车工程、无人驾驶等相关专业用作教材,也可作为智能网联、无人驾驶相关的企业人才培训、技能认证等的培训用书。

图书在版编目(CIP)数据

智能网联汽车 ROS 实战入门 / 曾子铭,深圳市元创兴科技有限公司编著. -- 北京:北京航空航天大学出版社,2021.8
 ISBN 978-7-5124-3512-4

Ⅰ. ①智… Ⅱ. ①曾… ②深… Ⅲ. ①汽车-智能通信网-高等职业教育-教材 Ⅳ. ①U463.67

中国版本图书馆 CIP 数据核字(2021)第 162156 号

版权所有,侵权必究。

智能网联汽车 ROS 实战入门
曾子铭
深圳市元创兴科技有限公司 编著
策划编辑 冯 颖 责任编辑 王 实

*

北京航空航天大学出版社出版发行

北京市海淀区学院路 37 号(邮编 100191) http://www.buaapress.com.cn
发行部电话:(010)82317024 传真:(010)82328026
读者信箱:goodtextbook@126.com 邮购电话:(010)82316936
北京宏伟双华印刷有限公司印装 各地书店经销

*

开本:787×1 092 1/16 印张:11.25 字数:288 千字
2021 年 8 月第 1 版 2021 年 8 月第 1 次印刷 印数:2 000 册
ISBN 978-7-5124-3512-4 定价:35.00 元

若本书有倒页、脱页、缺页等印装质量问题,请与本社发行部联系调换。联系电话:(010)82317024

前　言

随着全球汽车保有量的快速增长,能源短缺、环境污染、交通拥堵、事故频发等社会问题日益突出,现已成为汽车行业可持续发展的制约因素。智能网联汽车被公认为是这些问题的有效解决方案,代表着未来汽车行业的发展方向,也是我国目前汽车工业转型的重要方向之一。

关于本书

智能网联汽车是智能汽车和车联网结合的产物,对智能网联汽车行业来说,自动驾驶必将成为终极目标。随着自动驾驶技术的成熟,自动驾驶必将全面取代人工驾驶,到那时大量的相关产业就需要懂得自动驾驶技术的人才。本书以实操的方式全面地介绍了自动驾驶汽车编程知识,而且配套了完备的教学资源,可帮助学生快速掌握自动驾驶车辆的基础原理以及结构组成,也可满足自动驾驶、智能网联汽车等相关课程的教学需求。通过对此类相关课程的学习,学生毕业后可从事与自动驾驶相关产业的售后甚至研发工作,为自动驾驶产业爆发式增长积蓄人才。

本书按照循序渐进、由浅入深的方式,介绍了ROS(机器人操作系统)的概念、ROS工程的搭建和配置、编写一个ROS节点,以及ROS通信机制等知识,相信它会是一本非常实用的实践教学用书。本书要求读者具备一定的C++语言以及Linux基础,没有安装ROS的小伙伴可以参照ROS官网上的安装步骤自行安装,本教程使用的环境为Ubuntu 16.04＋ROS kinetic。ROS安装参考链接:http://wiki.ros.org/cn/kinetic/Installation/Ubuntu。

本书的可操作性较强,既可参照书中的内容编程和学习,也可快速更好地掌握智能网联汽车知识,快速查到自己所关心的问题的相关解决办法。

关于作者

曾子铭　毕业于英国威尔士大学,博士后,深圳市高层次人才,深圳市创业导师,现为深圳职业技术学院汽车与交通学院教师,研究方向为无人驾驶、人工智能和计算机视觉。曾主持或参与国家及省市级课题12项,作为第一作者发表SCI、EI检索论文20篇,获发明专利3项、实用新型专利6项。

深圳市元创兴科技有限公司 创立于2006年,是一家从事智能控制、机器人及人工智能教育方案的国家高新技术企业,致力于智能控制、机器人和人工智能等相关技术的研发,为企业和高校提供设备定制、实验室建设等全方位服务。

致 谢

本书在编写过程中,引入企业实战资源和图片,在此特向其作者和图片制作者表示诚挚的谢意。

由于时间仓促,书中不足之处在所难免,恳盼读者给予指正。希望本书的出版能对普及智能网联汽车知识,发展智能网联汽车起到积极的引导和促进作用。

作 者
2021年5月

目　　录

第1章　智能网联汽车 ROS 认知 ·· 1
1.1　什么是 ROS ··· 3
1.2　ROS 在智能网联汽车开发中的角色 ································· 3
1.3　ROS 的总体目标和特点 ··· 4
1.3.1　ROS 设计的总体目标 ·· 4
1.3.2　ROS 的特点 ·· 5
课后练习 ·· 8

第2章　认识 ROS 工程 ··· 9
2.1　ROS 环境搭建与配置 ·· 9
2.1.1　将 Linux 系统安装到移动 U 盘 ································ 9
2.1.2　在 Linux 系统中安装 ROS ···································· 18
2.2　ROS 工程的基本概念 ··· 20
2.3　Catkin 工作空间 ··· 21
2.3.1　新建一个工作空间 ··· 21
2.3.2　编译工作空间 ··· 21
2.3.3　工作空间的结构 ·· 23
2.4　Package 功能包 ·· 24
2.4.1　Package 的结构 ·· 24
2.4.2　CMakeLists.txt 文件 ··· 25
2.4.3　package.xml 文件 ··· 26
2.4.4　创建 Package 功能包 ··· 26
2.4.5　Package 相关命令 ··· 27
课后练习 ··· 28

第3章　编写一个 ROS 节点 ··· 30
3.1　节点及节点管理器 ·· 30
3.2　roscpp ··· 30
3.3　节点编写过程 ·· 32
3.3.1　初始化节点 ·· 32
3.3.2　关闭节点 ··· 32
3.3.3　NodeHandle 常用成员函数 ···································· 32
3.3.4　节点编写 ··· 33
3.3.5　CMakeLists.txt 文件的修改 ··································· 34
3.3.6　运行生成的节点 ·· 35
3.3.7　rosnode 命令 ·· 36

3.4 头文件引用	37
3.4.1 引用当前包头文件	37
3.4.2 引用同一工作空间内其他软件包的头文件	39
3.5 第三方库文件引用	41
3.6 launch 文件编写	44
3.6.1 \<launch\> 标签	45
3.6.2 \<node\> 标签	45
3.6.3 \<include\> 标签	48
3.6.4 \<param\> 标签	51
3.6.5 \<arg\> 标签	53
3.6.6 \<group\> 标签	53
课后练习	56
第4章 ROS 通信机制——话题	**57**
4.1 话题通信原理	57
4.2 话题通信示例	58
4.3 rqt_graph 和 rqt_plot 命令的使用	61
4.4 rostopic 命令的使用	62
4.5 代码解析	64
4.5.1 third_pkg.cpp 源码分析	64
4.5.2 subscrbe.cpp 源码分析	65
4.6 理解自定义消息类型	66
4.6.1 什么是消息	66
4.6.2 rosmsg 命令	66
4.7 创建自定义消息类型	67
课后练习	70
第5章 ROS 通信机制——服务	**72**
5.1 认识服务基本概念	72
5.2 编写 ROS 服务示例	73
课后练习	76
第6章 参数服务器	**78**
6.1 roscpp 中的 rosparam	78
6.1.1 getParam()	78
6.1.2 Param()	78
6.1.3 setParam()	79
6.1.4 deleteParam()	79
6.1.5 hasParam()	79
6.2 通过 launch 加载参数	79
6.3 rosparam 命令	82
6.4 动态参数调节	83

课后练习 …………………………………………………………………………………… 88

第7章 ROS通信机制——动作 ………………………………………………………… 89
7.1 动作简介 …………………………………………………………………………… 89
7.2 动作文件规范 ……………………………………………………………………… 90
7.3 编写一个动作示例 ………………………………………………………………… 91
课后练习 …………………………………………………………………………………… 96

第8章 什么是tf ……………………………………………………………………… 98
8.1 tf介绍 ……………………………………………………………………………… 98
8.2 tf示例 ……………………………………………………………………………… 98
8.2.1 示例运行 ……………………………………………………………………… 98
8.2.2 tf命令工具 …………………………………………………………………… 99
8.2.3 tf中的消息 ………………………………………………………………… 101
8.3 tf的C++接口 …………………………………………………………………… 102
8.3.1 数据类型 …………………………………………………………………… 102
8.3.2 数据转换 …………………………………………………………………… 103
8.3.3 tf类 ………………………………………………………………………… 107
课后练习 ………………………………………………………………………………… 109

第9章 ROS车型机器人建模 ………………………………………………………… 110
9.1 机器人组成架构 ………………………………………………………………… 110
9.1.1 常见执行机构 ……………………………………………………………… 111
9.1.2 驱动系统实现 ……………………………………………………………… 112
9.1.3 传感器系统 ………………………………………………………………… 112
9.1.4 控制系统实现 ……………………………………………………………… 114
9.2 URDF描述语言解释 …………………………………………………………… 114
9.3 创建URDF模型 ………………………………………………………………… 117
9.3.1 创建机器人描述功能包 …………………………………………………… 117
9.3.2 创建URDF模型 …………………………………………………………… 118
9.3.3 URDF解析 ………………………………………………………………… 121
9.3.4 在rviz中显示机器人模型 ………………………………………………… 122
9.4 改进URDF模型 ………………………………………………………………… 125
9.4.1 使用xacro优化URDF模型 ……………………………………………… 125
9.4.2 引用xacro文件 …………………………………………………………… 129
9.4.3 显示优化后的模型 ………………………………………………………… 129
9.5 添加传感器 ……………………………………………………………………… 131
9.5.1 添加摄像头 ………………………………………………………………… 131
9.5.2 添加激光雷达 ……………………………………………………………… 132
9.6 gazebo仿真 ……………………………………………………………………… 134
9.6.1 给base_link添加惯性、碰撞及gazebo属性 …………………………… 134
9.6.2 在gazebo中显示机器人模型 …………………………………………… 138

  9.6.3 gazebo 常用插件 ································· 139
 9.7 sw2urdf 工具 ·· 156
 课后练习 ··· 157

第 10 章 常见运动学解算 ·································· 159
 10.1 两轮差动模型 ·· 159
 10.2 三轮全向运动模型 ···································· 161
 10.3 四轮全驱滑动运动模型 ································ 163
 10.4 四轮全向运动模型 ···································· 165
 课后练习 ··· 169

参考文献 ··· 171

第 1 章 智能网联汽车 ROS 认知

智能网联汽车 ICV(Intelligent Connected Vehicle),是车联网与智能汽车的有机联合,搭载了先进的车载传感器、控制器、执行器等装置,并融合现代通信与网络技术来实现车与人、路、云端等信息交换共享。智能网联汽车具备复杂的环境感知、智能决策、协同控制等功能,可带来"安全、舒适、节能、高效"的行驶体验。随着新一代信息技术与汽车产业的深度融合,智能网联汽车正成为各国争相抢占的制高点。近年来,汽车产业从"电动化"向"智能化""网联化"升级,催生了自动驾驶、车联网等全新的行业。在人工智能、物联网的持续升温下,我国也十分重视智能网联汽车的发展。未来几年,在政策和行业趋势的指引下,智能网联汽车将迎来蓬勃发展期,也将成为我国汽车产业实现转型升级的一大关键性节点。

机场、港口等场所,以及快速公交车和产业园区通勤出行、智能物流配送、智能环卫等限定场景的优先规模化应用,为自动驾驶商用车企业带来发展先机。2018 年年底,工信部发布了《车联网(智能网联汽车)产业发展行动计划》(以下简称《行动计划》),明确表示将加大对车联网(智能网联汽车)行业的政策扶持力度,充分彰显了国家对于智能网联汽车行业的高度重视。《行动计划》明确以 2020 年为时间节点,分两个阶段实现智能网联汽车行业高质量发展的目标。第一阶段,到 2020 年,实现车联网(智能网联汽车)产业跨行业融合并取得突破,具备高级别自动驾驶功能的智能网联汽车实现特定场景规模应用,车联网用户渗透率达到 30% 以上,智能道路基础设施水平明显提升。第二阶段,2020 年后,技术创新、标准体系、基础设施、应用服务和安全保障体系将全面建成,高级别自动驾驶功能的智能网联汽车和 5G-V2X 逐步实现规模化商业应用,"人-车-路-云"实现高度协同。

智能网联汽车的核心是自动驾驶/无人驾驶,如图 1.1 所示。自动驾驶汽车(Autonomous vehicles;Self-piloting automobile)又称无人驾驶汽车、电脑驾驶汽车或轮式移动机器人,是一种通过电脑系统实现无人驾驶的智能汽车,并不是所有的智能网联汽车都可以实现自动驾驶,但自动驾驶汽车必须要借助智能网联技术,三者间存在着一种包

图 1.1 智能网联汽车

含关系。智能网联汽车利用移动网络实现人车交互,就像一个装了轮子会跑的电脑一样,至于能实现什么功能,则取决于系统里安装的软件:安装浏览器可以上网,安装播放器可以看视频,安装地图可以查导航,同样,安装智能驾驶软件可以实现自动驾驶。目前很多车辆内置了类似自动泊车、定速巡航等功能,已经投入市场且相对成熟,但上述操作有限制性使用条件且需要驾驶员时刻紧盯驾驶状况,距离真正的"自动"与"无人"还有很大的距离。

自动驾驶/无人驾驶汽车实际属于机器人的研究范畴。综上所述,"机器人"的核心概念和功能,就是代替人从事一些人类无法完成或"不愿意"做的事情的一种自动化机构。人是一个

高度复杂的智能体,如果机器人要取代人完成一些工作,则机器人也必然是一个高度复杂的智能体。

机器人工程有以下特点:

(1) 综合性

首先机器人是一个系统工程,它涉及机械、电子、控制、通信、软件等诸多学科,如图 1.2 所示。

(2) 实践性

机器人工程专业是一个实践性极强的综合类工科专业,从机器人设计,到信息处理,再到控制决策,无一不需要实践的验证才有意义。

(3) 复杂性

机器人作为一个新兴的综合智能体在经过多年

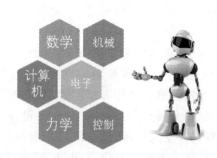

图 1.2　机器人工程是一个综合学科

的理论沉淀后走向工程化,本身就是多学科的综合,又处在工科领域的领先地位,是集软件、硬件、算法于一体的复杂综合体。

(4) 智能性

机器人存在的意义在于取代人类从事艰苦、危险和繁杂的工作,本质上是"替"人工作,而人类是高级的智慧性生物,因此机器人发展的终极目标就是智能。机器人是人工智能应用的重要领域。

因此,开发一个机器人需要花很大的功夫,需要设计机械结构、画电路板、写驱动程序、设计通信架构、组装集成、调试以及编写各种感知决策和控制算法,每一个任务都需要花费大量的时间。因此像电影《钢铁侠》中那样,仅靠一个人的力量造出一个动力超强的人形机器人机甲是不可能的。图 1.3 所示为智能机器人开发的知识架构。

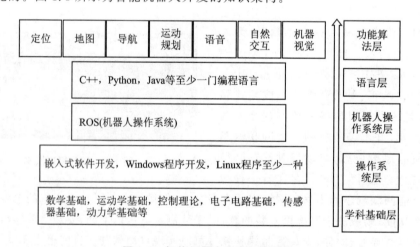

图 1.3　智能机器人开发知识架构

由于机器人开发难度大、从业门槛高以及缺乏开发标准等因素严重阻碍了机器人行业的发展。因此,当前急需一个标准的开发方法、一个高效统一的软件架构、一套辅助开发的工具来帮助广大的机器人开发者发挥自己的才智,构建机器人工程的"高楼大厦"。

1.1 什么是ROS

ROS是一个适用于机器人编程的框架,这个框架把原本松散的零部件耦合在一起,为它们提供通信架构。ROS虽然叫做操作系统,但并非Windows、Mac那种通常意义的操作系统,它只是连接了操作系统和你所开发的ROS应用程序,所以它也算是一个中间件,在基于ROS的应用程序之间建立起沟通的桥梁,是运行在基本操作系统上的环境,在这个环境中,机器人的感知、决策、控制算法可以更好地组织和运行。以上几个关键词(框架、中间件、操作系统、运行的环境)都可以用来描述ROS的特性,作为初学者不必深究这些概念,随着越来越多地使用ROS,就能够体会到它的作用:

- 一种模块化软件通信机制;
- 一系列先进的算法:SLAM、ORK、Moveit!;
- 一系列开源工具:3D显示、坐标转换、实时监控等;
- 一款跨平台的开发环境;
- 一个最活跃的机器人开发交流平台。

1.2 ROS在智能网联汽车开发中的角色

图1.4所示为ROS在智能机器人开发中的角色。一个机器人系统一般由机器人本体、操作系统以及大脑组成。

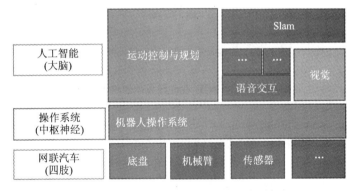

图1.4 ROS在智能机器人开发中的角色

机器人本体包括执行机构及感知系统。执行机构通常包含运动底盘、机械手臂、机械手等机械或电气执行单元。感知系统一般由视觉、语音、激光、IMU等传感器组成。

机器人大脑主要由环境信息处理以及机器人行为决策控制等一系列算法模块组成。常见机器人的功能有:环境建模与定位、自然交互、运动控制与规划、计算机视觉等。

操作系统提供算法单元之间的连接及算法与执行机构之间的连接,相当于人类的中枢神经,即"小脑"。ROS在机器人开发中扮演的角色就是"小脑"。图1.5所示为ROS在机器人开发中的应用。

ROS在无人驾驶、工业领域、教育领域都有很广泛的应用。图1.6所示为ROS在各行业中的应用。

图 1.5　ROS 在机器人开发中的应用

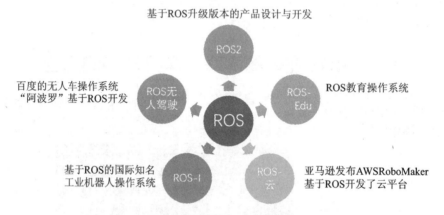

图 1.6　ROS 在各行业中的应用

1.3　ROS 的总体目标和特点

1.3.1　ROS 设计的总体目标

如图 1.7 所示，左图和右图是制作自行车的两种模式，左图是传统模式，大家都在各自"闭门造车"，"重复造轮子"，效率低下。右图是现代模式，设计、制造、集成，分工明确，效率得到了很大提高。

图 1.7 中右图的现代模式效率明显高于左图的传统模式，现代模式之所以能够分工明确，重要因素在于：统一的标准。自行车的各个部件和工序都有明确的标准，这样各个厂商都可以按照该标准执行，不用担心自己所造的"轮子"跟别人的"车架"不配套。

当前机器人行业，特别是智能机器人行业存在的问题就是缺乏统一的标准，没有统一的硬件架构、软件架构和操作系统，甚至没有统一的开发语言。这严重阻碍了技术和人才的交流，导致一个机器人功能的开发和维护只能由机器人本身的生产厂商来做。几乎不能像手机一样

图 1.7　自行车制造

由第三方开发人员来开发大量优质的、能适用于大部分机器人的应用。ROS 的诞生解决了这一问题，ROS 设计的总体目标就是提高代码复用率，为机器人开发提供统一的标准。

1.3.2　ROS 的特点

ROS 的特点如图 1.8 所示。

1. 点对点

所谓点对点，可以理解为：代码解耦、分布式计算、多机协作。

代码解耦：如图 1.9 中的 Robot1，ROS 允许将整个工程的代码解耦设计成 N 个独立的可执行程序。如 Robot1 中的 camera 节点负责相机的数据采集，vision 节点负责相机图像的处理。至于节点之间的通信和数据交换则由 ROS 的通信框架来解决。代码解耦带来两个好处：首先，单个节点的代码量减少，方便

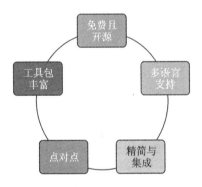

图 1.8　ROS 的特点

维护。其次，代码的复用率大大提高，比如在 Robot2 中如果也要用到 Robot1 同样的相机采集图像，则可以直接使用 Robot1 中的 camera 节点。

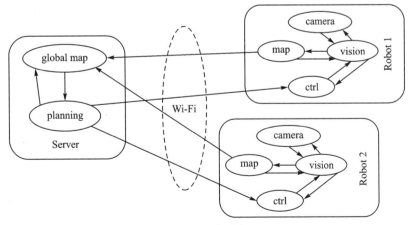

图 1.9　点对点

分布式计算:假设 Robot1 的控制器运算能力不足,ROS 允许将 Robot1 中的这些节点部署在多个控制器上,只需要保证这些控制器在同一个网络上即可,至于通信问题 ROS 会解决。比如:可以将比较消耗资源的 camera 和 vision 部署在额外的控制器上。这样,可将这两块控制器视为"同一块"控制器。

多机协作:如图 1.9 所示,ROS 同样允许将多个节点部署在多个机器人上,这些机器人可以协作共同完成任务。比如共同构建"一张地图",共同使用"一张地图"等。

2. 多语言支持

ROS 为机器人开发者提供了不同语言的编程接口,比如 C++接口叫做 roscpp,Python 接口叫做 rospy,Java 接口叫做 rosjava 等。表 1.1 所列为目前 ROS 支持的开发语言客户端。

表 1.1 ROS 支持的开发语言客户端

客户端库	介绍
roscpp	ROS 的 C++库,是目前应用最广泛的 ROS 客户端库,执行效率高
rospy	ROS 的 Python 库,开发效率高,通常用在对运行时间没有太多要求的场合,例如配置、初始化等操作
roslisp	ROS 的 LISP 库
roscs	Mono/.NET. 库,可用任何 Mono/.NET 语言,包括 C#、Iron Python、IronRuby 等
rosgo	ROS Go 语言库
rosjava	ROS Java 语言库
rosnodejs	Java script 客户端库
…	…

目前最常用的有 roscpp 和 rospy,而其余的语言版本基本还是测试版。

3. 精简与集成

ROS 利用了很多现在已经存在的开源项目的代码(见图 1.10),比如说从 Player 项目中借鉴了驱动、运动控制和仿真方面的代码,从 OpenCV 中借鉴了视觉算法方面的代码,从 OpenRAVE 借鉴了规划算法的内容,还有很多其他项目。在每一个实例中,ROS 都用来显示多种配置选项以及与各软件之间进行数据通信,同时也对它们进行微小的包装和改动。ROS 可以不断地从社区维护中升级,包括从其他的软件库、应用补丁中升级 ROS 的源代码。

图 1.10 ROS 集成了众多开源项目和库

4. 工具包丰富

为了管理复杂的 ROS 软件框架,我们利用大量的小工具去编译和运行多种多样的 ROS 组件,从而设计成了内核,而不是构建一个庞大的开发和运行环境。

这些工具担任了各种各样的任务,例如,组织源代码的结构,获取和设置配置参数,形象化端对端的拓扑连接,测量频带使用宽度,生动地描绘信息数据,自动生成文档,等等。尽管我们已经测试通过了全局时钟和控制器模块的记录器的核心服务,但我们还是希望能把所有的代码模块化。我们相信,效率上的损失远远是稳定性和管理的复杂性上无法弥补的。ROS 常用工具见图 1.11。

rviz 信息 3 维显示工具

gazebo 3 维仿真工具

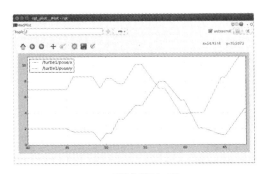
rqt 可视化调试工具

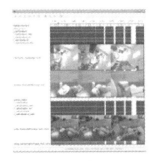
rosbag 数据记录与回放工具

图 1.11　ROS 常用工具

5. 免费且开源

ROS 所有的源代码都是公开发布的(见图 1.12),这必定促进 ROS 软件各层次的调试,不断地改正错误。虽然 Microsoft Robotics Studio 和 Webots 这样的非开源软件也有很多值得赞美的属性,但我们认为一个开源的平台是无可替代的。当硬件和各层次的软件同时设计和调试时,这一点是尤其真实的。

图 1.12　免费且开源

ROS 以分布式的关系遵循 BSD 许可。BSD 许可是一个给予使用者很大自由的协议,允许各种商业和非商业的工程进行开发。使用者可以进行二次开发和修改源代码,也可以将修改后的代码作为开源或者专有软件再发布。

总之,ROS 提供了一套通信机制允许代码解耦提高代码的复用率,提供了一系列的开发工具辅助开发,集成了大量的机器人应用功能,建立了一个机器人开发的生态系统,促使机器人开发标准化。

课后练习

一、选择题

(1) [单选]智能网联汽车要实现的最终目标是(　　)。

　　(A) 实现汽车高度自动化/无人驾驶

　　(B) 实现智能化

　　(C) 实现网联化

　　(D) 实现电动化

(2) [单选]机器人操作系统的全称是(　　)。

　　(A) React Operating System

　　(B) Router OperatingSytstem

　　(C) Request of Service

　　(D) Robot Operating System

(3) [单选]下列不属于 ROS 的特点是(　　)。

　　(A) 开源

　　(B) 分布式架构

　　(C) 强实时性

　　(D) 模块化

二、简答题

(1) 智能网联汽车的定义是什么?

(2) 举例说明 ROS 在智能网联汽车开发中的应用。

第 2 章 认识 ROS 工程

本章主要介绍 ROS 的工程结构,也就是 ROS 的文件系统结构。要学会建立一个 ROS 工程,首先要认识 ROS 工程,了解它们的组织架构,从根本上熟悉 ROS 项目的组织形式,了解各个文件的功能和作用,才能正确地进行开发和编程。本章的主要内容有,介绍 Catkin 的编译系统,Catkin 工作空间的创建和结构,Package 软件包的创建和结构,介绍 CMakeLists.txt 文件、package.xml 文件以及其他常见文件,并系统地梳理了 ROS 文件空间的结构。这些对于我们学习和开发 ROS 起着重要的作用。

2.1 ROS 环境搭建与配置

2.1.1 将 Linux 系统安装到移动 U 盘

ROS 环境搭建需要准备的材料包括:一个移动 U 盘(容量不少于 8 GB,最好是固态 U 盘)、VMWare 虚拟机(如:Vmware 15.5.0)、Linux 系统镜像文件(如:ubuntu 16.04 LTS)。

环境搭建与配置步骤:

① 安装 VMWare 虚拟机(如:Vmware 15.5.0)。安装完成后,打开 VMWare 软件(右击图标,用管理员身份运行),然后创建一个虚拟机(见图 2.1),选择配置的类型,然后单击"下一步"按钮。

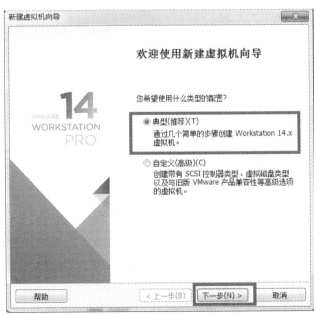

图 2.1 创建一个虚拟机(选择类型配置)

选择稍后安装操作系统(见图2.2),单击"下一步"按钮。

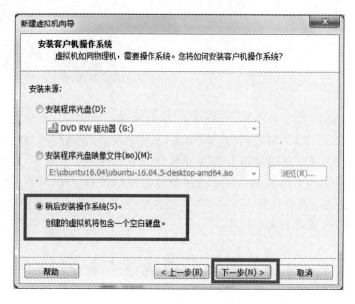

图2.2 创建一个虚拟机(安装来源)

客户机操作系统选择 Linux(L),版本选择 Ubuntu 64位(见图2.3),然后单击"下一步"按钮。

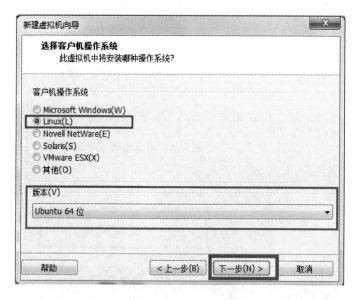

图2.3 创建一个虚拟机(选择客户机操作系统和版本)

输入虚拟机名称并设置保存路径(见图2.4),然后单击"下一步"按钮。
指定磁盘容量,选择将虚拟磁盘拆分成多个文件(见图2.5),然后单击"下一步"按钮。
最后单击"完成"按钮,完成虚拟机的创建(见图2.6)。
② 编辑创建的虚拟机,选择编辑虚拟机设置(见图2.7)。
添加 Linux 系统镜像文件(如:ubuntu-16.04.4-desktop-amd64.iso)路径(见图2.8)。

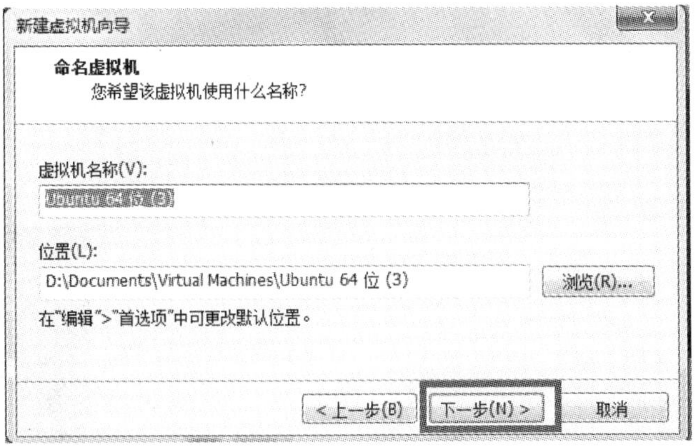

图 2.4　创建一个虚拟机(输入虚拟机名称和位置)

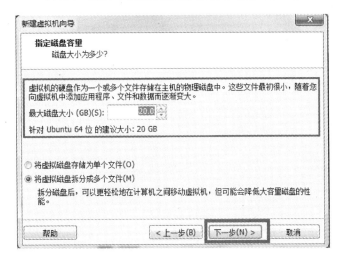

图 2.5　创建一个虚拟机(指定磁盘容量)

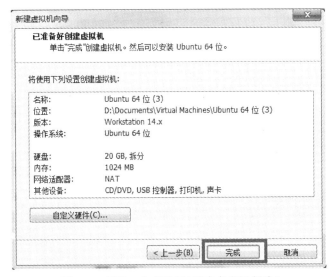

图 2.6　创建一个虚拟机(完成虚拟机创建)

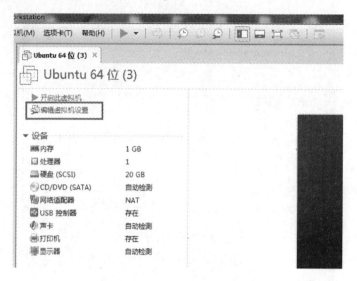

图 2.7　编辑虚拟机(选择编辑虚拟机设置)

图 2.8　编辑虚拟机(硬件设置)

将原来的"硬盘"删除(见图 2.9)。

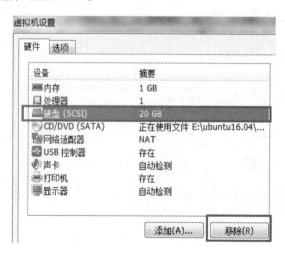

图 2.9　编辑虚拟机(删除原硬盘)

然后创建一个新的"硬盘",添加自己的移动设备(U盘)(见图2.10)。

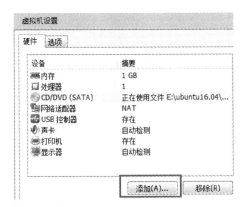

图 2.10　编辑虚拟机(添加 U 盘)

选择硬件类型为硬盘(见图2.11)。

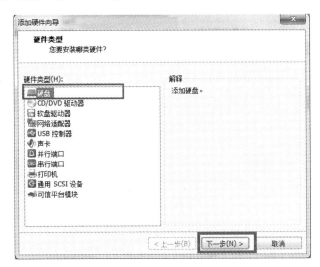

图 2.11　编辑虚拟机(选择硬件类型)

选择虚拟磁盘类型为 SCSI(S)(见图2.12)。

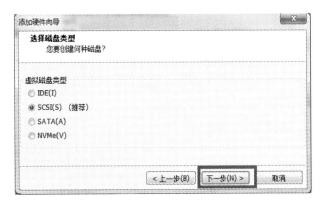

图 2.12　编辑虚拟机(选择虚拟磁盘类型)

选择使用物理磁盘(见图2.13)。

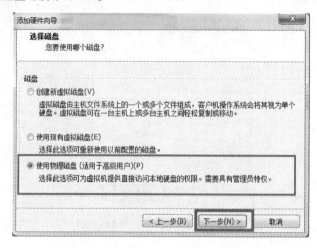

图2.13　编辑虚拟机(选择使用物理磁盘)

选择本地磁盘设备(见图2.14)。

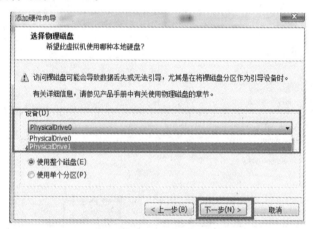

图2.14　编辑虚拟机(选择本地磁盘设备)

指定此虚拟磁盘文件将存储选定物理磁盘的分区信息(见图2.15),然后单击"完成"按钮。

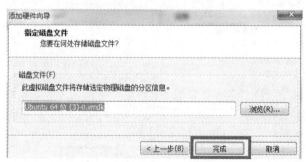

图2.15　编辑虚拟机(指定磁盘文件)

完成后的效果如图 2.16 所示,单击"确定"按钮。

图 2.16 编辑虚拟机(总体配置)

③ 编辑虚拟机选项,选择打开电源时进入固件(见图 2.17)。

图 2.17 编辑虚拟机选项

进入 BIOS 后,将光标移动到 Boot 选项,选择启动项 CD-ROM Drive 选项(见图 2.18,按组合键 Shift+"+")移动到第一项,效果如图 2.19 所示。

完成修改设置后,按 F10 保存 BIOS 设置,然后进入安装系统环节,系统安装完成后关闭虚拟机,最后将安装了 Linux 系统的移动 U 盘拔出。

④ 启动 U 盘中的 Linux 系统。

电脑开机时进入 BIOS(比如,若电脑是华硕笔记本(N550JV)则按 F2,其他牌子的电脑需要上网查一下)设置启动项(见图 2.20)。

在 Boot 选项中将 Boot Option #1 设置为 U 盘启动(见图 2.21,如 aigo U391 固态 U 盘)。

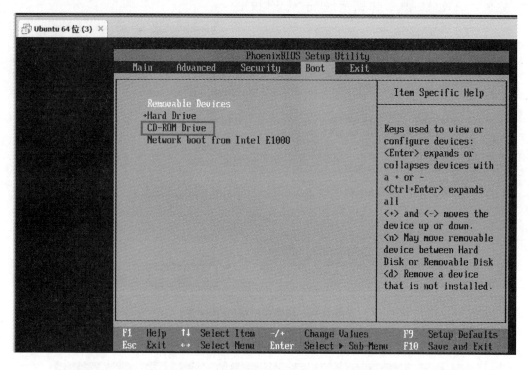

图 2.18　Boot 选项设置

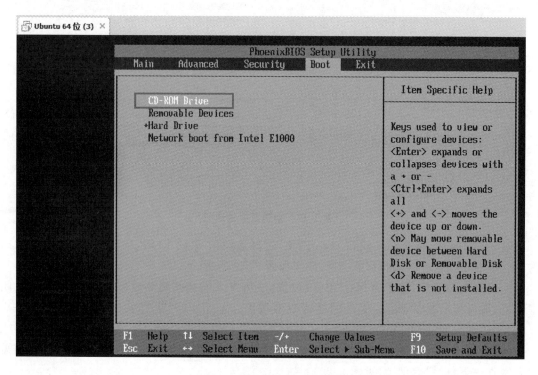

图 2.19　Boot 选项设置(已设置好启动顺序)

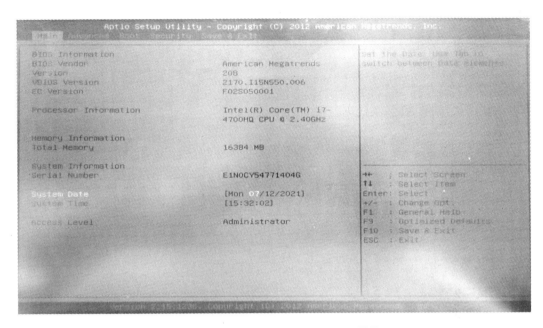

图 2.20　华硕笔记本(N550JV)BIOS 设置界面

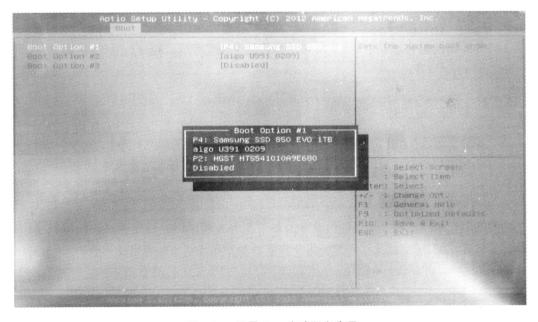

图 2.21　设置 Boot 启动顺序选项

设置完成后进入 BIOS 设置启动项，可以看到 Boot Option #1 已经设置为 U 盘启动，效果如图 2.22 所示。

按快捷键 F10 保存设置并退出 BIOS，然后系统会自动重启电脑，启动 Linux 系统(Ubuntu)。

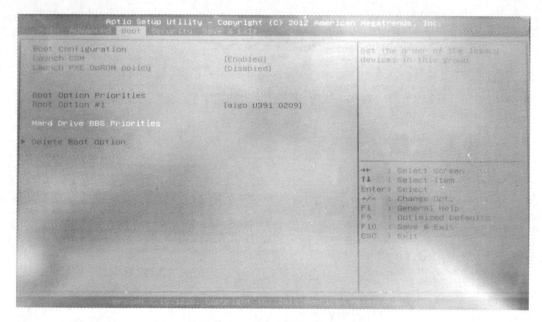

图 2.22　完成启动顺序设置

2.1.2　在 Linux 系统中安装 ROS

1. 设置 sources.list

为 Ubuntu 的包管理器增加源,设置计算机接受来自于 packages.ros.org 的软件,设置 sources.list 命令如下:

sudo sh - c 'echo "deb http://packages.ros.org/ros/ubuntu $(lsb_release - sc) main" > /etc/apt/sources.list.d/ros - latest.list'

这一步会根据 Ubuntu Linux 发行版本的不同,添加不同的源。Ubuntu 的版本通过 lsb_release - sc 获得。添加正确的软件库之后,操作系统就知道去哪里下载程序,并根据命令自动安装软件。

2. 设置密钥

为确认源代码的正确性并且确保没有人在未经所有者授权的情况下修改任何程序代码,当添加完软件库时,需添加软件库的密钥并将其添加到操作系统的可信任列表中,设置密钥的命令如下:

sudo apt - key adv -- keyserver hkp://ha.pool.sks - keyservers.net:80 -- recv - key 421C365BD9FF1F717815A3895523BAEEB01FA116

如果在连接密钥服务器时遇到问题,可以尝试在上面的命令中用 hkp://pgp.mit.edu:80 或 hkp://keyserver.ubuntu.com:80 来替换。

3. 安　装

首先,需要确保包管理器的索引已经更新至最新,进行更新操作的命令如下:

sudo apt - get update

ROS 中有非常多不同的库和工具,官方提供了 4 种默认的配置来安装 ROS,也可以独立安装 ROS 包,推荐采用桌面完整安装方式(安装内容包括 ROS、rqt、rviz、机器人通用库、2D/3D 仿真器、导航及 2D/3D 感知),桌面完整安装的命令如下:

```
sudo apt-get install ros-kinetic-desktop-full
```

4. 安装功能包的系统依赖

在使用 ROS 之前,需要先初始化 rosdep。rosdep 可以用于编译源码和运行 ROS 核心组件,并简单地安装系统依赖,初始化 rosdep 的命令如下:

```
sudo rosdep init
```

如果初始化失败,打开 hosts 文件,并进行如下操作:

- 打开 hosts 文件,命令如下:

```
sudo gedit /etc/hosts
```

- 在文件末尾添加,命令如下:

```
151.101.84.133    raw.githubusercontent.com
```

- 保存后退出,再尝试 rosdep 初始化命令,初始化 rosdep 成功后,安装和更新功能包的系统依赖,命令如下:

```
rosdep update
```

5. 配置 ROS 环境

如果在每次启动一个新的终端时,ROS 环境变量都能自动地添加进你的 bash 会话环境中,这是非常方便的,可以通过如下命令来实现:

```
echo "source /opt/ros/kinetic/setup.bash" >> ~/.bashrc
source ~/.bashrc
```

6. 构建包所需的依赖

到这一步,已经安装好运行核心 ROS 包的所有东西。要创建和管理你自己的 ROS 工作空间,还有单独发布的许多工具(比如,rosinstall 是一个常用的命令行工具,使你可以通过一个命令为 ROS 包简单地下载许多源码树)。要安装这个工具及其他的依赖以构建 ROS 包,则运行如下命令:

```
sudo apt-get install python-rosinstall python-rosinstall-generator python-wstool build-essential
```

完整的 ROS 安装完成之后,可以通过 roscore 和 turtlesim 来对安装做一个简单的测试。运行如下命令:

```
roscore
rosrun turtlesim turtlesim_node
rosrun turtlesim turtle_teleop_key
```

为了能够将 pip 升级到最新版,运行如下命令:

```
python -m pip install -U pip
```

7. 安装程序运行需要的包

sudo apt－get install ros－kinetic－map－server ros－kinetic－teleop－twist－keyboard ros－kinetic－slam－gmapping ros－kinetic－navigation＊ros－kinetic－move－base＊ros－kinetic－amcl ros－kinetic－dwa－local－planner ros－kinetic－teb－local－planner ros－kinetic－navfn ros－kinetic－global－planner tree ros－inetic－arbotix

2.2 ROS 工程的基本概念

如图 2.23 所示为一个 ROS 工程的组织形式与内容。

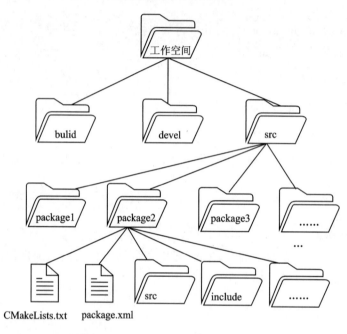

图 2.23 ROS 工程的组织形式与内容

图中各部分说明如下：

工作空间　包含该机器人工程的所有内容以及功能。一般一个机器人工程需要一个工作空间。

build　存放工程编译的中间文件。

devel　存放工程编译生成的可执行代码、编译生成的库文件以及自动生成的头文件等。

src　存放该工程的功能包。

package　功能包是 ROS 工程中的基本单元，一般一个单一机器人功能建立一个功能包。

功能包清单（package.xml）　用于记录功能包的基本信息，包含作者信息、许可信息、依赖选项和编译标志等。每个功能包都包含一个名为 package.xml 的功能包清单。

CMakeLists.txt　功能包的编译规则。每个功能包都必须包含一个 CMakeLists.txt 文件。

src　存储功能包源代码。

include　存储功能包头文件。

2.3 Catkin 工作空间

Catkin 工作空间是创建、修改、编译 Catkin 软件包的目录。Catkin 工作空间,直观地形容就是一个仓库,里面装载着 ROS 的各种项目工程,便于系统组织管理调用。在可视化图形界面里是一个文件夹。我们自己写的 ROS 代码通常就放在工作空间里,本节介绍 Catkin 工作空间的结构。

2.3.1 新建一个工作空间

我们要在计算机上创建一个初始的 ros_workspace/ 路径,这也是 Catkin 工作空间结构的最高层级。输入下列指令,完成初始创建:

```
$ mkdir -p ~/ros_workspace/src
```

上面代码直接创建了第二层级的文件夹 src,这也是存放 ROS 软件包的地方。

2.3.2 编译工作空间

对于源代码包,只有编译才能在系统上运行。而 Linux 下的编译器有 gcc、g++,随着源文件的增加,直接用 gcc/g++ 命令的方式显得效率低下,人们开始用 Makefile 来进行编译。然而随着工程体量的增大,Makefile 也不能满足需求,于是便出现了 CMake 工具。CMake 是 Make 工具的生成器,是更高层的工具,它简化了编译构建过程,能够管理大型项目,具有良好的扩展性。对于 ROS 这样大体量的平台来说,就采用 CMake,并且 ROS 对 CMake 进行了扩展,于是便有了 Catkin 编译系统(见图 2.24)。

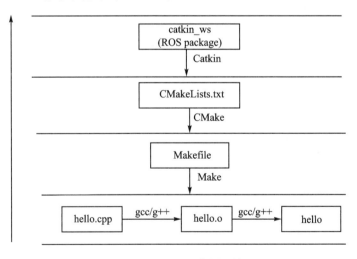

图 2.24 Catkin 编译系统

Catkin 是基于 CMake 的编译构建系统,具有以下特点:

Catkin 沿用了包管理的传统,像 find_package() 基础结构,pkg-config 扩展了 CMake,例如:

- 软件包编译后无需安装就可使用;

- 自动生成 find_package() 代码、pkg-config 文件；
- 解决了多个软件包构建顺序问题。

一个 Catkin 的软件包（package）必须包括两个文件：

① package.xml 包括了 package 的描述信息：name，description，version，maintainer(s)，license，opt. authors，url's，dependencies，plugins，等等。

② CMakeLists.txt 构建 package 所需的 CMake 文件：
- 调用 Catkin 的函数/宏；
- 解析 package.xml；
- 找到其他依赖的 Catkin 软件包；
- 将本软件包添加到环境变量。

Catkin 的的工作原理

Catkin 编译的工作流程如下：

① 在工作空间 catkin_ws/src/ 下递归地查找其中每一个 ROS 的 package。

② package 中会有 package.xml 和 CMakeLists.txt 文件，Catkin（CMake）编译系统依据 CMakeLists.txt 文件生成 Makefiles（放在 catkin_ws/build/）。

③ Make 刚刚生成的 Makefiles 等文件，编译链接生成可执行文件（放在 catkin_ws/devel）。

也就是说，Catkin 就是将 CMake 与 Make 指令做了一个封装，用来完成整个编译过程的工具。Catkin 比较突出的优点主要是：操作更加简单，一次配置，多次使用跨依赖项目编译。

要用 Catkin 编译一个工程或软件包，只需要用 catkin_make 指令。一般当我们写完代码，执行一次 catkin_make 进行编译，系统会自动完成编译和链接过程，构建生成目标文件。编译的一般性流程如下：

```
$ cd ~/ros_workspace/ #切换到工作空间目录
$ catkin_make #编译 ROS 工作空间
```

注意：Catkin 编译之前需要回到工作空间目录，catkin_make 在其他路径下编译不会成功。编译完成后，如果有新的目标文件产生（原来没有），那么一般要紧接着使用 source 刷新配置环境，使得系统能够找到刚才编译生成的 ROS 可执行文件。这个细节如果遗漏，将导致后面出现可执行文件无法打开等错误。

```
$ source ~/ros_workspace/devel/setup.bash #刷新坏境
```

上述代码操作刷新配置的环境变量只在当前环境变量中有效。如果想要该环境变量在所有的终端中都有效，则可使用以下方法：

```
$ echo "source ~/ros_workspace/devel/setup.sh" >> ~/.bashrc
$ source ~/.bashrc #使生效
```

catkin_make 命令也有一些可选参数，例：

```
catkin_make [args]
 -h, --help                              //帮助信息
 -C DIRECTORY, --directory DIRECTORY
 工作空间的路径（默认为 '.'）
 --source SOURCE                         //src 的路径（默认为 'workspace_base/src'）
 --build BUILD                           //build 的路径（默认为 'workspace_base/build'）
```

```
--use-ninja                          //用 ninja 取代 Make
--use-nmake                          //用 nmake 取代 Make
--force-cmake                        //强制 CMake,即使已经过 CMake
--no-color                           //禁止彩色输出(只对 catkin_make 和 CMake 生效)
--pkg PKG                            //[PKG...] 只对某个 PKG 进行 Make
--only-pkg-with-deps                 //ONLY_PKG_WITH_DEPS [ONLY_PKG_WITH_DEPS...]
```
将指定的 package 列入白名单 CATKIN_WHITELIST_PACKAGES,
只编译白名单里的 package。该环境变量存在于 CMakeCache.txt。
```
--cmake-args [CMAKE_ARGS [CMAKE_ARGS...]]   //传给 CMake 的参数
--make-args [MAKE_ARGS [MAKE_ARGS...]]      //传给 Make 的参数
--override-build-tool-check                  //用来覆盖由不同编译工具产生的错误
```

2.3.3 工作空间的结构

初看起来 Catkin 工作空间极其复杂,其实不然,Catkin 工作空间的结构其实非常清晰。具体的 Catkin 工作空间结构图如下。

在工作空间下用 tree 命令,可显示 Catkin 工作空间的结构。

```
$ cd ~/ros_workspace
$ sudo apt install tree
$ tree
```

结果如下:

```
── build
│   ├── catkin
│   │   └── catkin_generated
│   │       └── version
│   │           └── package.cmake
│   ├──
│   ......
│   ├── catkin_make.cache
│   ├── CMakeCache.txt
│   ├── CMakeFiles
│   │   └──
│   ......
├── devel
│   ├── env.sh
│   ├── lib
│   ├── setup.bash
│   ├── setup.sh
│   ├── _setup_util.py
│   └── setup.zsh
├── src
└── CMakeLists.txt -> /opt/ros/kinetic/share/catkin/cmake/toplevel.cmake
```

通过 tree 命令可以看到 Catkin 工作空间的结构,它包括了 src、build、devel 三个路径,在有些编译选项下也可能包括其他路径。但这三个文件夹是 Catkin 编译系统默认的。它们的具体作用如下:

src/: ROS 的 Catkin 软件包(源代码包)。

build/：Catkin(CMake)的缓存信息和中间文件。
devel/：生成目标文件（包括头文件、动态链接库、静态链接库、可执行文件等）、环境变量。

在编译过程中，它们的工作流程如图 2.25 所示。

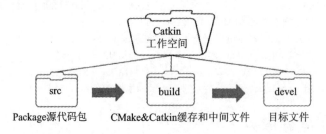

图 2.25　编译过程

后两个路径由 Catkin 系统自动生成和管理，这在日常的开发中一般不会涉及，而主要用到的是 src 文件夹，我们写的 ROS 程序、网上下载的 ROS 源代码包都存放在这里。在编译时，Catkin 编译系统会递归地查找和编译 src/ 下的每一个源代码包。因此，也可以把几个源代码包放到同一个文件夹下，如图 2.26 所示。

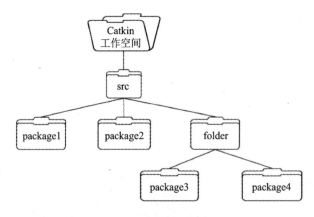

图 2.26　功能包组织形式

2.4　Package 功能包

ROS 中的 Package 的定义更加具体，它不仅是 Linux 上的软件包，更是 Catkin 编译的基本单元，我们调用 catkin_make 编译的对象就是一个个 ROS 的 Package，也就是说，任何 ROS 程序只有组织成 Package 才能编译。所以，Package 也是 ROS 源代码存放的地方，任何 ROS 的代码无论是 C++ 还是 Python 都要放到 Package 中，这样才能正常地编译和运行。一个 Package 可以编译多个目标文件（ROS 可执行程序、动态静态库、头文件等）。

2.4.1　Package 的结构

一个 Package 下常见的文件、路径有：

```
├── CMakeLists.txt  # package 的编译规则(必须)
├── package.xml     # package 的描述信息(必须)
├── src/            # 源代码文件
├── include/        # C++头文件
├── scripts/        # 可执行脚本
├── msg/            # 自定义消息
├── srv/            # 自定义服务
├── models/         # 3D 模型文件
├── urdf/           # urdf 文件
├── launch/         # launch 文件
```

其中,定义 Package 的是 CMakeLists.txt 和 package.xml,这两个文件是 Package 中必不可少的。Catkin 编译系统在编译前,首先就要解析这两个文件。这两个文件就定义了一个 Package。

- CMakeLists.txt:定义 Package 的包名、依赖、源文件、目标文件等编译规则,是 Package 不可缺少的成分;
- package.xml:描述 Package 的包名、版本号、作者、依赖等信息,是 Package 不可缺少的成分;
- src/:存放 ROS 的源代码,包括 C++的源码(.cpp)以及 Python 的 module(.py);
- include/:存放 C++源码对应的头文件;
- scripts/:存放可执行脚本,例如 shell 脚本(.sh)、Python 脚本(.py);
- msg/:存放自定义格式的消息(.msg);
- srv/:存放自定义格式的服务(.srv);
- models/:存放机器人或仿真场景的 3D 模型(.sda,.stl,.dae 等);
- urdf/:存放机器人的模型描述(.urdf 或.xacro);
- launch/:存放 launch 文件(.launch 或.xml)。

通常 ROS 文件组织都是按照以上的形式,这是约定俗成的命名习惯,建议遵守。以上路径中,只有 CMakeLists.txt 和 package.xml 是必须的,其余路径根据软件包是否需要来决定。

2.4.2 CMakeLists.txt 文件

CMakeLists.txt 原本是 CMake 编译系统的规则文件,而 Catkin 编译系统基本沿用了 CMake 的编译风格,只是针对 ROS 工程添加了一些宏定义。所以在写法上,Catkin 的 CMakeLists.txt 与 CMake 基本一致。

这个文件直接规定了这个 Package 要依赖哪些 Package,要编译生成哪些目标,如何编译等流程。所以,CMakeLists.txt 非常重要,它指定了由源码到目标文件的规则,Catkin 编译系统在工作时首先会找到每个 Package 下的 CMakeLists.txt,然后按照规则来编译构建。

CMakeLists.txt 的基本语法是按照 CMake 的,而 Catkin 在其中加入了少量的宏,总体的结构如下:

```
cmake_minimum_required()  # CMake 的版本号
project()                 # 项目名称
find_package()            # 找到编译需要的其他 CMake/Catkin Package
catkin_python_setup()     # Catkin 新加宏,打开 Catkin 的 Python Module 的支持
```

```
add_message_files()  #Catkin新加宏,添加自定义Message/Service/Action文件
add_service_files()
add_action_files()
generate_message()  #Catkin新加宏,生成不同语言版本的msg/srv/action接口
catkin_package()  #Catkin新加宏,生成当前Package的CMake配置,供依赖本包的其他软件包调用
add_library()  #生成库
add_executable()  #生成可执行的二进制文件
add_dependencies()  #定义目标文件依赖于其他目标文件,确保其他目标已被构建
target_link_libraries()  #链接
catkin_add_gtest()  #Catkin新加宏,生成测试
install()  #安装至本机
```

如果从未接触过 CMake 的语法,请阅读《CMake 实践》(https://github.com/Akagi201/learning-cmake/blob/master/docs/cmake-practice.pdf)。掌握 CMake 语法对于理解 ROS 工程很有帮助。

2.4.3　package.xml 文件

package.xml 也是 catkin 的 Package 必备文件,它是这个软件包的描述文件,在较早的 ROS 版本(rosbuild 编译系统)中,这个文件叫做 manifest.xml,用于描述 Pacakge 的基本信息。如果在网上看到一些 ROS 项目里包含着 manifest.xml,那么它多半是 hydro 版本之前的项目。

pacakge.xml 包含了 Package 的名称、版本号、内容描述、维护人员、软件许可、编译构建工具、编译依赖、运行依赖等信息。实际上,rospack find、rosdep 等命令之所以能快速定位和分析出 Package 的依赖项信息,就是直接读取了每一个 Pacakge 中的 package.xml 文件。它为用户提供了快速了解一个 Pacakge 的渠道。

pacakge.xml 遵循 xml 标签文本的写法,通常包含以下标签:

```
<pacakge> 根标记文件
<name> 包名
<version> 版本号
<description> 内容描述
<maintainer> 维护者
<license> 软件许可证
<buildtool_depend> 编译构建工具,通常为 Catkin
<depend> 指定依赖项为编译、导出、运行需要的依赖,最常用
<build_depend> 编译依赖项
<build_export_depend> 导出依赖项
<exec_depend> 运行依赖项
<test_depend> 测试用例依赖项
<doc_depend> 文档依赖项
```

2.4.4　创建 Package 功能包

创建一个 Package 需要在 ros_workspace/src 下,使用 catkin_create_pkg 命令,用法如下:

```
catkin_create_pkg package depends
```

其中,package 是包名,depends 是依赖的包名,可以依赖多个软件包。

例如,新建一个 Package 叫做 test_pkg ,依赖 roscpp、rospy、std_msgs(常用依赖)。

注意:ROS 包的命名遵循一个命名规范,只允许使用小写字母、数字和下画线,而且首字符必须是一个小写字母。一些 ROS 工具,包括 Catkin,不支持那些不遵循此命名规范的包。

输入以下命令创建第一个功能包:

```
$ cd ~/ros_workspace/src  # 切换到功能包创建目录
$ catkin_create_pkg test_pkg roscpp rospy std_msgs
# 创建名字为 test_pkg 的功能包,依赖了 roscpp,rospy,std_msgs
```

这样就会在当前路径下新建 test_pkg 软件包,包括:

```
├── CMakeLists.txt
├── include
│   └── test_pkg
├── package.xml
└── src
```

catkin_create_pkg 帮你完成了软件包的初始化,填充好了 CMakeLists.txt 和 package.xml,并且将依赖项填进了这两个文件中。

2.4.5 Package 相关命令

1. rospack

rospack 是对 Package 管理的工具,命令的用法如表 2.1 所列。

表 2.1　rospack 命令

rospack 命令	作　用
rospack help	显示 rospack 的用法
rospack list	列出本机所有 Package
rospack depends [package]	显示 Package 的依赖包
rospack find [package]	定位某个 Package
rospack profile	刷新所有 Package 的位置记录

2. roscd

roscd 命令类似于 Linux 系统的 cd,改进之处在于 roscd 可以直接 cd 到 ROS 的软件包。roscd 命令的用法如表 2.2 所列。

表 2.2　roscd 命令

roscd 命令	作　用
roscd [pacakge]	cd 到 ROS Package 所在路径

3. rosls

rosls 也可以视为 Linux 指令 ls 的改进版,可以直接 ls ROS 软件包的内容。rosls 命令的用法如表 2.3 所列。

表 2.3　rosls 命令

rosls 命令	作　用
rosls [pacakge]	列出 ROS Package 下的文件

4. rosdep

rosdep 是用于管理 ROS Package 依赖项的命令行工具,用法如表 2.4 所列。

表 2.4　rosdep 命令

rosdep 命令	作　用
rosdep check [pacakge]	检查 Package 的依赖是否满足
rosdep install [pacakge]	安装 Pacakge 的依赖
rosdep db	生成和显示依赖数据库
rosdep init	初始化 /etc/ros/rosdep 中的源
rosdep keys	检查 Package 的依赖是否满足
rosdep update	更新本地的 rosdep 数据库

一个较常使用的命令是

rosdep install -- from - paths src -- ignore - src -- rosdistro = kinetic - y

用于安装工作空间中 src 路径下所有 Package 的依赖项(由 pacakge.xml 文件指定)。

课后练习

一、选择题

(1) [单选]ROS Kinetic 最佳适配的 Linux 版本是(　　)。

　　(A) CentOS 7

　　(B) Ubuntu 14.04

　　(C) Ubuntu 16.04

　　(D) Ubuntu 18.04

(2) [单选]ROS 官方二进制包可以通过的命令安装(假定 Kinetic 版本)是(　　)。

　　(A) sudo apt - get install ROS_kinetic_desktop_full

　　(B) sudo apt - get install ROS - Kinetic - desktop - full

　　(C) sudo apt - get install ros_kinetic_desktop_full

　　(D) sudo apt - get install ros - kinetic - desktop - full

(3) [单选]启动 ROS Master 的命令是(　　)。

　　(A) roscore

　　(B) rosmaster

　　(C) rosMaster

　　(D) roslaunch

(4)[单选]在默认情况下,catkin_make 生成的 ROS 可执行文件所放的路径是(   )。
    (A) catkin_ws/src
    (B) catkin_ws/
    (C) catkin_ws/devel
    (D) catkin_ws/build
(5)[多选]下列属于 ROS 的发行版本是(   )。
    (A) Indigo
    (B) Jade
    (C) Xenial
    (D) Kinetic

二、判断题
roscd、rosls 指令后面都可以直接加包名,作用分别是跳转到软件包路径下、列出软件包中的内容。(   )
    (A) 正确
    (B) 错误

第 3 章

编写一个 ROS 节点

3.1 节点及节点管理器

在 ROS 的世界里，最小的进程单元就是节点（node）。一个软件包里可以有多个可执行文件，可执行文件在运行之后就成了一个进程（process），这个进程在 ROS 中叫做节点。从程序角度来说，节点就是一个可执行文件（通常为 C++编译生成的可执行文件、Python 脚本）被执行，加载到了内存之中；从功能角度来说，通常一个节点负责着机器人的某一项单独的功能。由于机器人的功能模块非常复杂，因此我们往往不会把所有功能都集中到一个节点上，而会采用分布的方式，把鸡蛋放到不同的篮子里。例如：用一个节点控制底盘轮子的运动，用一个节点驱动摄像头获取图像，用一个节点驱动激光雷达，用一个节点根据传感器信息进行路径规划……这样做可以降低程序发生崩溃的可能性。试想一下，如果把所有功能都写到一个程序中，模块间的通信、异常处理将会很麻烦。

由于机器人的元器件很多，功能庞大，因此实际运行时往往会运行众多的节点，负责感知世界、控制运动、决策和计算等功能。那么，如何合理地进行调配、管理这些节点呢？这就要利用 ROS 提供的节点管理器（master），节点管理器在整个网络通信架构里相当于管理中心，管理着各个节点。节点首先在节点管理器处进行注册，之后节点管理器会将该节点纳入整个 ROS 程序中。节点之间的通信也是先由节点管理器进行"牵线"，才能两两地进行点对点通信。当 ROS 程序启动时，第一步首先启动节点管理器，由节点管理器处理依次启动节点。

下面介绍如何用 C++编写一个节点。

3.2 roscpp

ROS 为机器人开发者提供了不同语言的编程接口，比如 C++接口叫做 roscpp，Python 接口叫做 rospy，Java 接口叫做 rosjava。尽管语言不同，但这些接口都可以用来创建话题、帮助和参数，实现 ROS 的通信功能。客户端库有点类似开发中的帮助类，把一些常用的基本功能进行了封装。

ROS 支持的客户端库如表 3.1 所列。

表 3.1 客户端库

客户端库	介　　绍
roscpp	ROS 的 C++库，是目前应用最广泛的 ROS 客户端库，执行效率高
rospy	ROS 的 Python 库，开发效率高，通常用在对运行时间没有太多要求的场合，例如配置、初始化等操作

续表 3.1

客户端库	介绍
roslisp	ROS 的 LISP 库
roscs	Mono/.NET.库,可用任何 Mono/.NET 语言,包括 C#,Iron Python,Iron Ruby 等
rosgo	ROS Go 语言库
rosjava	ROS Java 语言库
rosnodejs	Javascript 客户端库
…	…

目前,最常用的只有 roscpp 和 rospy,而其他的语言版本基本还是测试版。

从开发客户端库的角度看,一个客户端库,至少要包括节点管理器注册、名称管理、消息收发等功能。这样才能给开发者提供对 ROS 通信架构进行配置的方法。

roscpp 位于 /opt/ros/kinetic 之下,用 C++ 实现了 ROS 通信。在 ROS 中,C++ 的代码是通过 Catkin 编译系统(扩展的 CMake)来进行编译构建的。可以简单地理解为把 roscpp 作为一个 C++ 的库,我们创建一个 CMake 工程,其中包括了 roscpp 等 ROS 的库,这样就可以在工程中使用 ROS 提供的函数了。

通常要调用 ROS 的 C++ 接口,首先就需要 #include <ros/ros.h>。

roscpp 的主要部分包括:

- ros::init()   解析传入的 ROS 参数,创建节点第一步需要用到的函数;
- ros::NodeHandle   与课题、帮助、参数等交互的公共接口;
- ros::master   包含从节点管理器查询信息的函数;
- ros::this_node   包含查询这个进程(node)的函数;
- ros::service   包含查询服务的函数;
- ros::param   包含查询参数服务器的函数,而不需要用到 NodeHandle;
- ros::names   包含处理 ROS 资源名称的函数,

具体可见 http://docs.ros.org/api/roscpp/html/index.html。

以上功能可以分为下面几类:

- Initialization and Shutdown   初始与关闭;
- Topics   话题;
- Services   服务;
- Parameter Server   参数服务器;
- Timers   定时器;
- NodeHandles   节点句柄;
- Callbacks and Spinning   回调和自旋(或者翻译为轮询);
- Logging   日志;
- Names and Node Information   名称管理;
- Time   时钟;
- Exception   异常。

看到这么多接口,千万别觉得复杂,因为在日常开发中并不会用到所有的功能,现在只需要对它们有一些印象,掌握几个比较常见和重要的用法就足够了。下面介绍关键的用法。

3.3 节点编写过程

介绍了roscpp的一些接口之后,下面来编写第一个ROS节点。

当执行一个ROS程序时,就被加载到了内存中,成了一个进程,在ROS里叫做节点。每一个ROS的节点尽管功能不同,但都有必不可少的一些步骤,比如初始化、销毁,需要通行的场景通常还需要节点的句柄。本节学习节点的一些最基本的操作。

3.3.1 初始化节点

对于一个C++编写的ROS程序,它区别于普通C++程序,是因为代码中做了两层工作:
① 调用了ros::init()函数,从而初始化节点的名称和其他信息,ROS程序一般都会以这种方式开始。
② 创建 ros::NodeHandle 对象,也就是节点的句柄,它可以用来创建Publisher、Subscriber以及做其他事情。

句柄(Handle)这个概念可以理解为一个"把手",握住了门把手,就可以很容易把整扇门拉开,而不必关心门是什么样子。NodeHandle就是对节点资源的描述,有了它就可以操作这个节点了,比如为程序提供服务、监听某个话题上的消息、访问和修改参数等。

3.3.2 关闭节点

通常要关闭一个节点可以直接在终端上按Ctrl+C组合键,系统会自动触发SIGINT句柄来关闭这个进程。也可以通过调用ros::shutdown()来手动关闭节点,但通常我们很少这样做。

以下是一个节点初始化、关闭的例子:

```
#include <ros/ros.h>
int main(int argc, char** argv)
{
ros::init(argc, argv, "your_node_name");
ros::NodeHandle nh;
//....节点功能
//....
ros::spin();//用于触发topic、service的响应队列
return 0;
}
```

这段代码是最常见的一个ROS程序的执行步骤,通常要启动节点,获取句柄,而关闭的工作系统会自动完成,如果有特殊需要也可以自定义。那么句柄可以用来做些什么呢?下面来看看NodeHandle常用的成员函数。

3.3.3 NodeHandle常用成员函数

NodeHandle是节点的句柄,用来对当前节点进行各种操作。在ROS中,NodeHandle是

一个定义好的类,通过 include <ros/ros.h>,可以创建这个类,以及使用它的成员函数。
NodeHandle 常用成员函数包括:

```
//创建话题的发布者
ros::Publisher advertise(const string &topic, uint32_t queue_size, bool latch = false);
//第一个参数是发布话题的名称
//第二个参数是消息队列的最大长度,如果发布的消息超过这个长度而没有被接收,那么旧的消息就会
//出队。通常设为一个较小的数即可
//第三个参数是是否锁存。某些话题并不是以某个频率发布,比如/map 这个话题,只有在初次订阅或者
//地图更新这两种情况下,/map 才会发布消息。这里就用到了锁存
//创建话题的订阅者
ros::Subscriber subscribe(const string &topic, uint32_t queue_size, void(*)(M));
//第一个参数是订阅话题的名称
//第二个参数是订阅队列的长度,如果收到的消息都没来得及处理,那么新消息入队,旧消息就会出队
//第三个参数是回调函数指针,指向回调函数来处理接收到的消息
//创建服务的服务对象,提供服务
ros::ServiceServer advertiseService(const string &service, bool(*srv_func)(Mreq &, Mres &));
//第一个参数是服务对象名称
//第二个参数是服务函数的指针,指向服务函数。指向的函数应该有两个参数,分别为接受请求和响应
//创建服务的客户端
ros::ServiceClient serviceClient(const string &service_name, bool persistent = false);
//第一个参数是服务对象名称
//第二个参数用于设置服务的连接是否持续,如果为 true,客户端将会保持与远程主机的连接,这样后
//续的请求会快一些。通常我们设为 flase
//查询某个参数的值
bool getParam(const string &key, std::string &s);
bool getParam (const std::string &key, double &d) const;
bool getParam (const std::string &key, int &i) const;
//从参数服务器上获取 key 对应的值,已重载了多个类型
//给参数赋值
void setParam (const std::string &key, const std::string &s) const;
void setParam (const std::string &key, const char * s) const;
void setParam (const std::string &key, int i) const;
//给 key 对应的 val 赋值,重载了多个类型的 val
```

可以看出,NodeHandle 对象在 ROS C++程序里非常重要,各种类型的通信都需要用 NodeHandle 来创建完成。后面我们会逐步学习如何使用这些函数。

3.3.4 节点编写

在第 2 章中我们建立了一个 ros_workspace 的工作空间,同时在工作空间创建了一个命名为 test_pkg 的功能包。本小节我们继续在 test_pkg 中创建一个节点。在 ros_workspace/src/test_pkg/src 路径下创建一个.cpp 文件命名为:test.cpp,并输入以下代码:

```
#include"ros/ros.h"
#include"std_msgs/String.h"
#include"sstream"

int main(int argc,char **argv)
{
    ros::init(argc,argv,"talker");
    ros::NodeHandle n;
    ros::Publisher chatter_pub = n.advertise <std_msgs::String>("chatter",1000);
```

```cpp
  ros::Rate loop_rate(10);
  int count = 0;

  while(ros::ok())
  {
    std_msgs::String msg;
    std::stringstream ss;

    ss << "hello world" << count;
    msg.data = ss.str();

    ROS_INFO("%s",msg.data.c_str());
    chatter_pub.publish(msg);

    ros::spinOnce();
    loop_rate.sleep();
    ++count;
  }
    return 0;
}
```

上述测试代码创建了一个话题,并向其中发布"hello world"。

3.3.5 CMakeLists.txt 文件的修改

在完成了节点源码编写之后,需要修改 test_pkg 包下的 CMakeLists.txt 文件(具体路径为:ros_workspace/src/test_pkg/CMakeLists.txt),将编辑的源码生成可执行文件。

需要修改以下两条指令:

add_executable(${PROJECT_NAME}_node src/test_pkg_node.cpp)

该条指令用于指定将该功能包 src 目录下的哪个源文件编译为可执行程序,其中:
${PROJECT_NAME}_node 为生成可执行程序的名字,该名字可以任意指定。
src/test_pkg_node.cpp 为编译要使用的源码的文件名。

target_link_libraries(${PROJECT_NAME}_node
${catkin_LIBRARIES}
)

该条指令用于指定所使用的一些链接库。其中:
${PROJECT_NAME}_node 是在 add_executable 中生成的可执行文件。如图 3.1 所示为修改完成的 CMakeLists.txt 文件。

当修改完 CMakeLists.txt 文件后就可以在工作空间的根目录下使用 catkin_make 进行编译了。

$ cd ~/ros_workspace
$ catkin_make

编译成功如图 3.2 所示。

编译生成的可执行文件位于:/home/reinovo/ros_workspace/devel/lib/test_pkg/路径下,命名为 test_pkg_node。大家可以自行查看是否有该文件生成。

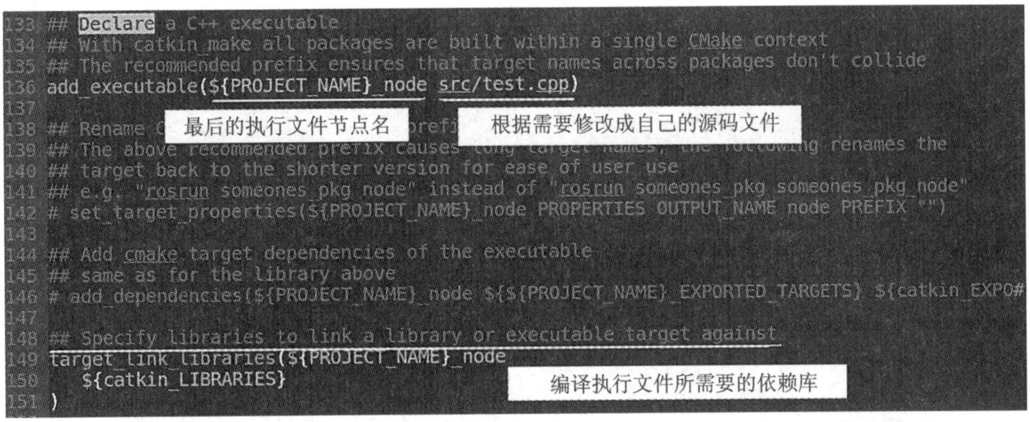

图 3.1 CMakeLists.txt 修改

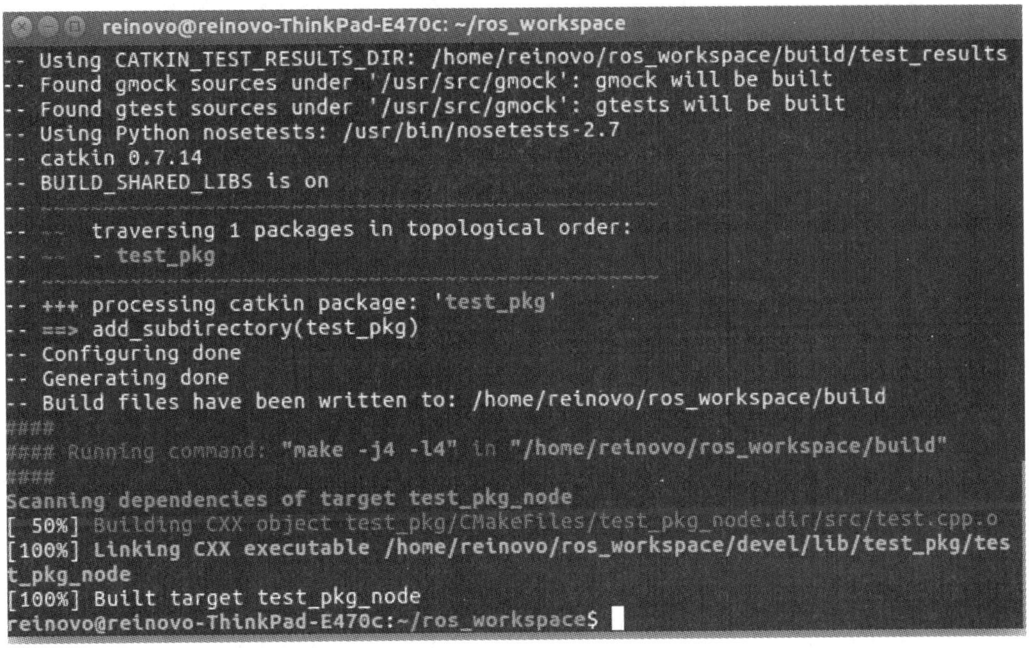

图 3.2 编译完成

3.3.6 运行生成的节点

打开一个终端输入以下命令：

$ roscore //打开节点管理器

打开节点管理器如图 3.3 所示。

重新打开一个终端输入以下命令并运行 test_pkg_node 节点：

$ rosrun test_pkg test_pkg_node

正确运行如图 3.4 所示。

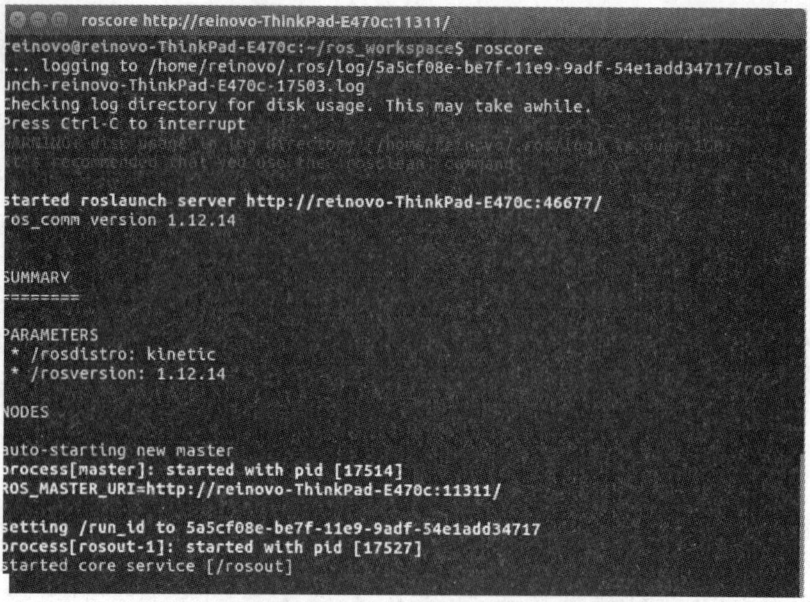

图 3.3 打开节点管理器

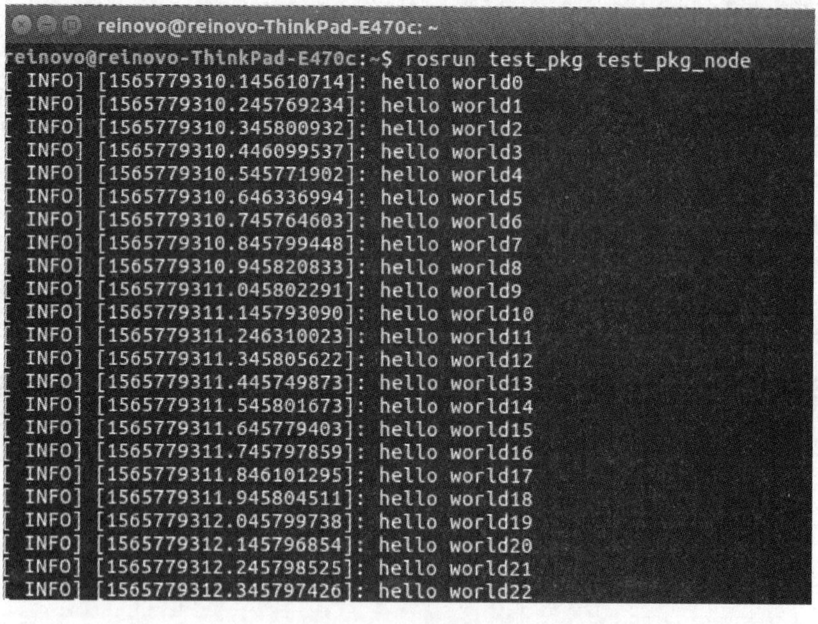

图 3.4 正确运行节点

注意：用 rosrun 运行节点之前一定要先运行 roscore，否则会报错。

3.3.7 rosnode 命令

ROS 提供了用于处理节点的命令 rosnode。

rosnode 命令的作用见表 3.2。

表 3.2 rosnode 命令

rosnode 命令	作用
rosnode list	列出当前运行的节点信息
rosnode info node.name	显示出节点的详细信息
rosnode kill node.name	结束某个节点
rosnode ping	测试连接节点
rosnode machine	列出在特定机器或列表机器上运行的节点
rosnode cleanup	清除不可到达节点的注册信息

3.4 头文件引用

3.4.1 引用当前包头文件

在编写代码时经常需要引用头文件,引用公用的头文件很容易,因为它们已经在标准库头文件路径中。但是如果要引用自定义的头文件就稍微麻烦点,我们首先查看软件包的目录结构,需要在正确的目录下创建头文件并修改 CMakeLists.txt 文件,这样才能正确编译,下面举例说明。

为了方便说明,先来创建一个头文件。

在路径~/ros_workspace/src/test_pkg/include/test_pkg 下创建 test_pkg.h。

打开终端,输入以下命令进入要创建头文件的目录:

$ cd ~/ros_workspace/src/test_pkg/include/test_pkg

输入以下命令创建头文件:

$ touch test_pkg.h

输入以下命令打开创建的头文件:

$ gedit test_pkg.h

输入以下代码:

```
#ifndef _TEST_PKG_
#define _TEST_PKG_
#define TEST_PKG_VER "1.0.0"
#define INIT_COUNT 100
int addTwoNum(int a,int b);
#endif
```

完成后保存关闭头文件,修改 test.cpp,代码如下:

```
#include "ros/ros.h"
#include "std_msgs/String.h"
#include "test_pkg/test_pkg.h" //自定义头文件
```

```cpp
#include <sstream>
int addTwoNum(int a,int b)
{
return a + b;
}//头文件函数实现部分
int main(int argc, char * argv[])
{
ros::init(argc, argv, "talker");//这个名字是给系统看的
ros::NodeHandle n;
ros::Publisher chatter_pub = n.advertise <std_msgs::String>("chatter", 1000);
ros::Rate loop_rate(10);
int count = INIT_COUNT;//调用了头文件中的宏定义
ROS_INFO("test_pkg version:%s",TEST_PKG_VER);
while (ros::ok())
{
std_msgs::String msg;
std::stringstream ss;
ss << "hello world " << count;
msg.data = ss.str();
ROS_INFO("%s", msg.data.c_str());chatter_pub.publish(msg);
ros::spinOnce();
loop_rate.sleep();
++count;
}
return 0;
}
```

源码修改完成之后需要修改 test_pkg 中的 CMakeLists.txt 文件。如图 3.5 所示(注意修改后要保存)。

```
include_directories(
    include
    ${catkin_INCLUDE_DIRS}
)
```

图 3.5 CMakeLists.txt 修改

此处修改是添加了一个头文件的包含路径:include。当然,此处的 include 是一个相对路径,指的是当前功能包 test_pkg 下的 include。这样,编译器在编译源码时会到 include 文件下查找我们自定义的头文件 test_pkg/test_pkg.h。

至此完成了如何在当前包下引用自定义头文件的代码修改以及配置修改。下面需要对源码重新编译。

$ cd ~/ros_workspace
$ catkin_make

编译完成之后运行:

$ roscore
$ rosrun test_pkg test_pkg_node

正确运行如图 3.6 所示,成功地输出了头文件中定义的版本号:version:1.0.0,并从 hello world 100 开始输出。

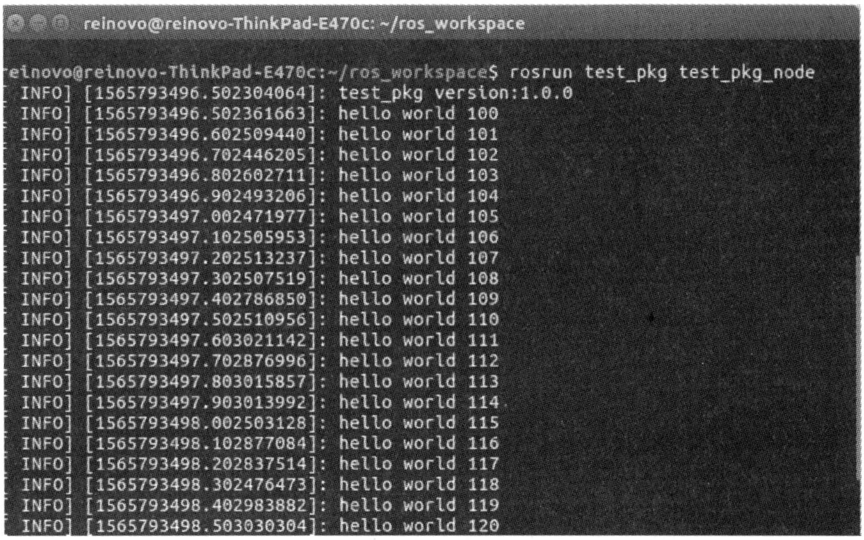

```
reinovo@reinovo-ThinkPad-E470c:~/ros_workspace$ rosrun test_pkg test_pkg_node
[ INFO] [1565793496.502304064]: test_pkg version:1.0.0
[ INFO] [1565793496.502361663]: hello world 100
[ INFO] [1565793496.602509440]: hello world 101
[ INFO] [1565793496.702446205]: hello world 102
[ INFO] [1565793496.802602711]: hello world 103
[ INFO] [1565793496.902493206]: hello world 104
[ INFO] [1565793497.002471977]: hello world 105
[ INFO] [1565793497.102505953]: hello world 106
[ INFO] [1565793497.202513237]: hello world 107
[ INFO] [1565793497.302507519]: hello world 108
[ INFO] [1565793497.402786850]: hello world 109
[ INFO] [1565793497.502510956]: hello world 110
[ INFO] [1565793497.603021142]: hello world 111
[ INFO] [1565793497.702876996]: hello world 112
[ INFO] [1565793497.803015857]: hello world 113
[ INFO] [1565793497.903013992]: hello world 114
[ INFO] [1565793498.002503128]: hello world 115
[ INFO] [1565793498.102877084]: hello world 116
[ INFO] [1565793498.202837514]: hello world 117
[ INFO] [1565793498.302476473]: hello world 118
[ INFO] [1565793498.402983882]: hello world 119
[ INFO] [1565793498.503030304]: hello world 120
```

图 3.6 成功引用了自定义头文件

3.4.2 引用同一工作空间内其他软件包的头文件

在一些情况下需要引用其他软件包中提供的函数或宏定义,这样可在一定程度上减少两个节点之间进行通信的话题个数,下面通过举例来进行说明。为了方便说明,需要在之前创建的工作空间中创建一个新的 package,并命名为 my_pkg,如图 3.7 所示。

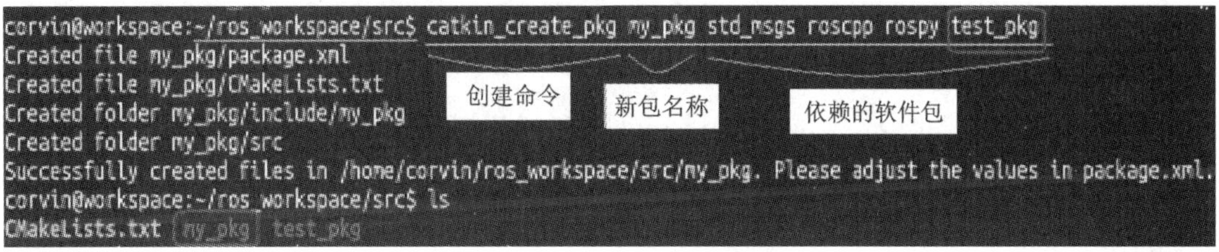

```
corvin@workspace:~/ros_workspace/src$ catkin_create_pkg my_pkg std_msgs roscpp rospy test_pkg
Created file my_pkg/package.xml
Created file my_pkg/CMakeLists.txt
Created folder my_pkg/include/my_pkg
Created folder my_pkg/src
Successfully created files in /home/corvin/ros_workspace/src/my_pkg. Please adjust the values in package.xml.
corvin@workspace:~/ros_workspace/src$ ls
CMakeLists.txt  my_pkg  test_pkg
```

图 3.7 创建 my_pkg 功能包

如何通过 catkin_create_pkg 命令来创建功能包在前面已经讲过,这里要注意的是创建 my_pkg 的目的是要引用 test_pkg 中的自定义头文件,因此这里创建 my_pkg 时要对 test_pkg 进行依赖。

在 my_pkg/src 路径下创建源码文件 my_pkg.cpp,写入以下代码:

```cpp
#include "ros/ros.h"
#include "std_msgs/String.h"
#include <sstream>
#include "test_pkg/test_pkg.h"//自定义头文件

int main(int argc, char *argv[])
{
    ros::init(argc, argv, "my_pkg");
    ros::NodeHandle n;
```

```cpp
ros::Publisher chatter_pub = n.advertise<std_msgs::String>("my_chatter",1000);
ros::Rate loop_rate(10);
int count = INIT_COUNT;//调用了头文件中的宏定义

ROS_INFO("test_pkg version:%s,init count:%d",TEST_PKG_VER,INIT_COUNT);
//将其他软件包头文件中声明的宏定义打印出来
while(ros::ok())
{
std_msgs::String msg;
std::stringstream ss;
ss << "my_pkg " << count;
msg.data = ss.str();
ROS_INFO("%s", msg.data.c_str());
chatter_pub.publish(msg);
ros::spinOnce();
loop_rate.sleep();
++count;
}
return 0;
}
```

在完成源码编写之后，首先是修改 my_pkg 的 CMakeLists.txt 文件。如图 3.8 所示，在 my_pkg 的 CMakeLists.txt 文件中指定了要编译的源码。

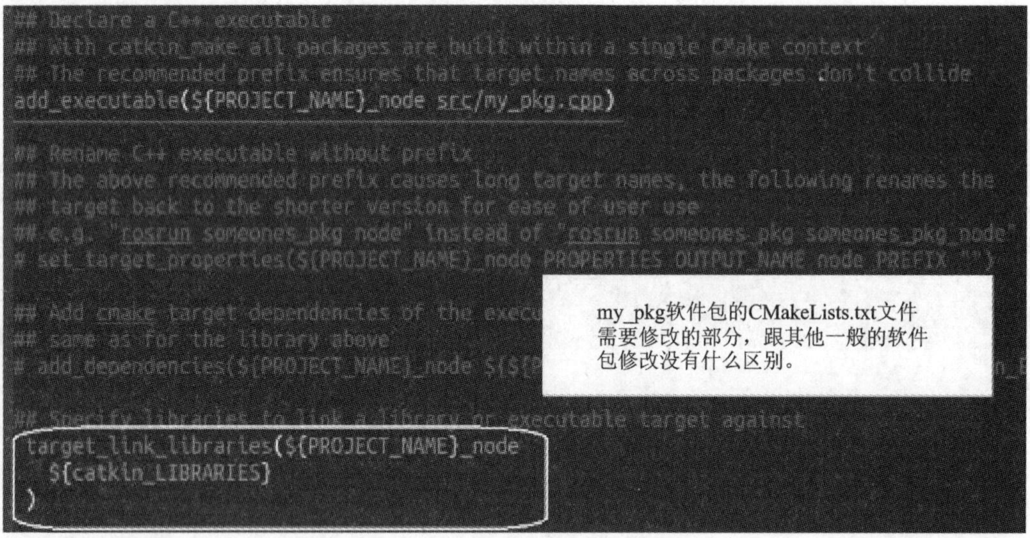

图 3.8 my_pkg CMakeLists.txt 修改

其次是修改 test_pkg 的 CMakeLists.txt 文件，如图 3.9 所示。做如下修改的主要目的是通知其他软件包当前软件包在 include 目录下含有自定义头文件。这样，其他包在引用头文件时就可以到这个目录下查找。

对工作空间进行编译：

```
$ cd ~/ros_workspace/
$ catkin_make
```

编译完成之后可以运行节点查看结果。

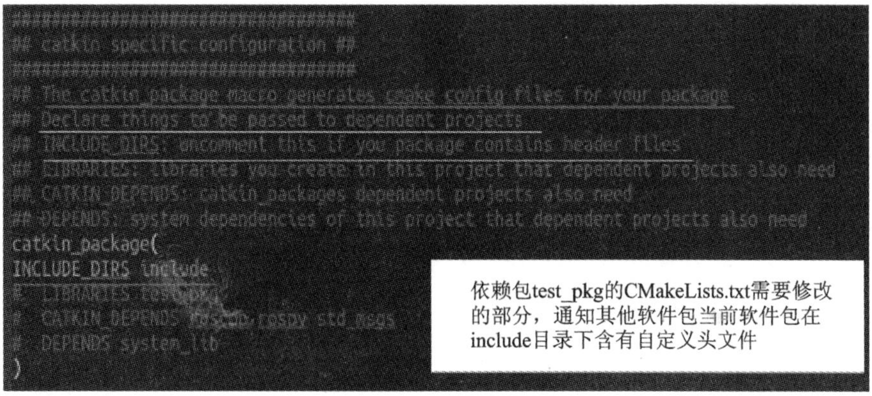

图 3.9 test_pkg CMakeLitst.txt 修改

```
$ roscore
$ rosrun my_pkg my_pkg_node
```

运行结果如图 3.10 所示，打印了头文件中定义的宏定义说明调用头文件成功。

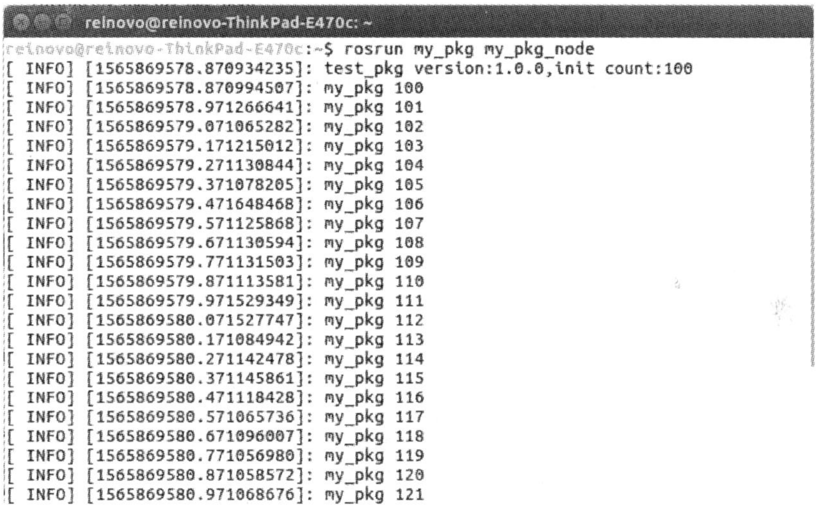

图 3.10 my_pkg_node 测试结果

大家也可以自行修改一下头文件中的宏定义，重新运行节点查看输出结果。

3.5 第三方库文件引用

我们经常在开发软件包时需要引入第三方的 .so 动态库，将其放到系统默认库路径中或者使用绝对路径，这样比较简单省事。但这样做软件包的移植性就会变差，当需要移植到其他机器上时，就要重新配置该软件包的依赖库路径。

为了更好地说明，我们手动创建一个 .so 动态库。

在任意目录下新建两个文件：multiply.cpp 和 multiply.h。

```
$ touch multiply.cpp multiply.h
```

打开 multiply.cpp 写入以下代码：

```cpp
#include "multiply.h"
int multiply(int a,int b)
{
return a * b;
}
```

打开 multiply.h 写入以下代码：

```cpp
#ifndef _MULTIPLY_H_
#define _MULTIPLY_H_
#ifdef __cpluscplus
extern "C"
{
#endif
int multiply(int a,int b);
#ifdef __cpluscplus
}
#endif
#endif
```

不难看出，multiply 提供了两个数相乘的功能。

接下来，准备开始将 multiply.cpp 编译成为动态库 libmultiply.so 供测试使用。在 multiply.h 以及 multiply.cpp 所在的路径下输入以下指令：

```
$ g++ -shared -fPIC -o libmultiply.so multiply.cpp
```

其中两个重要参数的意义如下：

-shared　告诉 gcc 要生成的是动态链接库；

-fPIC　告诉 gcc 生成的代码是非位置依赖的，方便用于动态链接。

在 my_pkg 包内创建 lib 目录，然后将生成的动态库 libmultiply.so 放到该目录下。同时，将动态库对应的头文件也放到当前软件包的 include 目录下，如图 3.11 所示。

图 3.11　my_pkg 文件结构图

修改 my_pkg.cpp 源码，在源码中调用第三方动态链接库中的函数。代码如下：

```cpp
#include "ros/ros.h"
```

编写一个 ROS 节点

```cpp
#include "std_msgs/String.h"
#include <sstream>
#include "test_pkg/test_pkg.h"
#include "my_pkg/multiply.h"   //动态库头文件
int main(int argc, char **argv)
{
    ros::init(argc,argv,"my_pkg");
    ros::NodeHandle n;
    ros::Publisher chatter_pub = n.advertise<std_msgs::String>("my_chatter", 1000);
    ros::Rate loop_rate(10);
    int count = multiply(30,5);//调用动态库中的方法
    ROS_INFO("test_pkg version:%s,init count:%d",TEST_PKG_VER,INIT_COUNT);
    while (ros::ok())
    {
        std_msgs::String msg;
        std::stringstream ss;
        ss << "my_pkg:" << count;
        msg.data = ss.str();
        ROS_INFO("%s", msg.data.c_str());
        chatter_pub.publish(msg);
        ros::spinOnce();
        loop_rate.sleep();
        ++count;
    }
    return 0;
}
```

接下来修改 my_pkg 的 CMakeLists.txt 文件，如图 3.12 所示。

```
## Specify additional locations of header files
## Your package locations should be listed before other locations
include_directories(
  include
  ${catkin_INCLUDE_DIRS}
)

link_directories(
lib
${catkin_LIB_DIRS}
)
```
包含自身头文件

link_directories()用于添加第三方库路径，catkin_LIB_DIRS表示环境变量，${catkin_LIB_DIRS}表示取环境变量的值。lib是相对于当前软件包所在路径的相对路径

```
## Declare a C++ library
# add_library(${PROJECT_NAME}
#   src/${PROJECT_NAME}/my_pkg.cpp
# )

## Add cmake target dependencies of the library
## as an example, code may need to be generated before libraries
## either from message generation or dynamic reconfigure
# add_dependencies(${PROJECT_NAME} ${${PROJECT_NAME}_EXPORTED_TARGETS} ${catkin_EXPORTED_TARGETS})

## Declare a C++ executable
## With catkin_make all packages are built within a single CMake context
## The recommended prefix ensures that target names across packages don't collide
add_executable(${PROJECT_NAME}_node src/my_pkg.cpp)

## Specify libraries to link a library or executable target against
target_link_libraries(${PROJECT_NAME}_node
   ${catkin_LIBRARIES}
   multiply
)
```
引用第三方动态库，在这里需要去掉lib前缀和后面的.so

图 3.12 修改后的 CMakeLists.txt

注意：link_directories 配置需要放置在 target_link_libraries 配置之前，否则会造成编译报错，其原因为需要通过 link_directories 找到添加的库文件。

编译：

```
$ cd ~/ros_workspace/
$ catkin_make
```

运行查看输出结果，如图 3.13 所示。

图 3.13　运行结果

3.6　launch 文件编写

在 ROS 中一个节点程序一般只能完成功能单一的任务，但是一个完整的 ROS 机器人一般由很多个节点程序同时运行、相互协作才能完成复杂的任务，因此要求在启动机器人时必须同时启动很多个节点程序，一般的 ROS 机器人由十几个节点程序组成，复杂的可能有几十个。

这就要求我们必须高效率地启动很多节点，而不是通过 rosrun 命令来依次启动十几个节点程序，launch 文件就是为解决这个需求而出现的。launch 文件是一个 xml 格式的脚本文件，把需要启动的节点写进 launch 文件中，并通过 roslaunch 工具来调用 launch 文件，执行这个脚本文件即可一次性启动所有的节点程序。

launch 文件同样也遵循 xml 格式规范，是一种标签文本，其格式包括以下标签：

```
<launch>     <!-- 根标签 -->
<node>       <!-- 需要启动的 node 及其参数 -->
<include>    <!-- 包含其他 launch -->
<machine>    <!-- 指定运行的机器 -->
<env-loader> <!-- 设置环境变量 -->
<param>      <!-- 定义参数到参数服务器 -->
<rosparam>   <!-- 启动 yaml 文件参数到参数服务器 -->
<arg>        <!-- 定义变量 -->
<remap>      <!-- 设定参数映射 -->
<group>      <!-- 设定命名空间 -->
```

</launch> <!-- 根标签 -->

官方参考链接：http://wiki.ros.org/roslaunch/XML。

下面创建一个功能包来进行说明：

```
$ cd ~/ros_workspace/src/
$ catkin_create_pkg robot_bringup
```

一般每一个机器人工程都包含一个命名为×××_bringup 的功能包，用来存放该机器人的各种启动文件。

robot_bringup 功能包创建成功后，在功能的目录下创建一个 launch 文件夹，并在 launch 文件夹下创建一个命名为 startup.launch 的文件。

```
$ cd ~/ros_workspace/src/robot_bringup/
$ mkdir launch
$ touch startup.launch
```

3.6.1 \<launch\> 标签

每个.xml 文件都必须包含一个根元素，这个根元素由一对 launch 标签定义：\<launch\> … \</launch\>。

打开 starup.launch 输入以下内容，如图 3.14 所示。

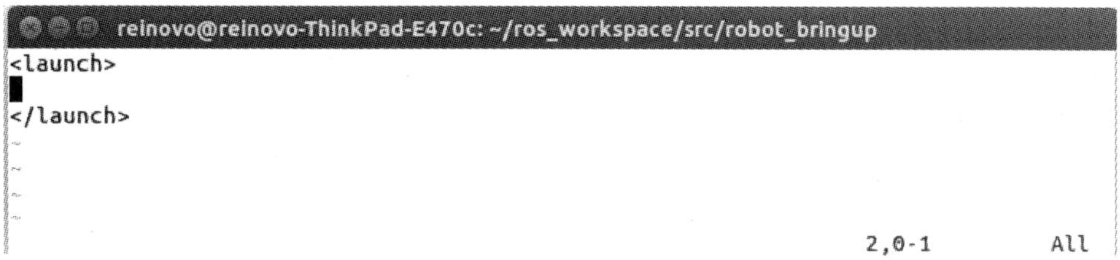

图 3.14 空的 launch 文件

3.6.2 \<node\> 标签

\<node\> 标签的上一级根标签为 \<launch\> 标签，用于启动一个 ROS 节点。启动一个节点的写法如下：

```
<node pkg = "package - name" type = "executable - name" name = "node - name" />
```

1. pkg，type，name，output 属性

\<node\> 标签下有若干属性，至少包含三个属性：pkg，type，name。pkg 属性指定了要运行的节点属于哪个功能包；type 是指要运行节点的可执行文件的名称；name 属性给节点指派了名称；它将覆盖任何通过调用 ros::int 来赋予节点的名称。

现在打开 startup.launch 文件，在里面运行 test_pkg_node 以及 my_pkg_node 两个节点。代码如下：

```
<launch>
<node pkg = "test_pkg" type = "test_pkg_node" name = "test_pkg_node" output = "screen"/>
<node pkg = "my_pkg" type = "my_pkg_node" name = "my_pkg_node" output = "screen"/>
```

```
</launch>
```

其中：

output="screen"属性可以将单个节点的标准输出到终端而不是存储在日志文件中。
编译工作空间并运行 startup.launch 文件。

```
$ cd ~/ros_workspace/
$ catkin_make
```

运行 launch 的命令格式如下：

roslaunch 包名称 launch 文件名称

运行 startup.launch 文件，结果如图 3.15 所示。

```
$ roslaunch robot_bringup startup.launch
```

图 3.15 运行结果

除了上述的 pkg, name, type, output 属性以外还有一些常用属性需要掌握。

2. 节点重生属性（respawn）

当 roslaunch 开启所有节点后，roslaunch 会监视每个节点，记录那些仍然活动的节点。对于每个节点，当其终止后，可以要求 roslaunch 重启该节点，并通过使用节点重生属性完成。对于一些即插即用的设备比如遥控手柄，重生属性非常有用。

修改 startup.launch 文件如下：

```
<launch>
<node pkg = "test_pkg" type = "test_pkg_node" name = "test_pkg_node" respawn = "true" output = "screen"/>
<node pkg = "my_pkg" type = "my_pkg_node" name = "my_pkg_node" output = "screen"/>
</launch>
```

保存并运行 launch 文件。

```
$ roslaunch robot_bringup startup.launch
```

输入命令，查看当前系统运行的节点，如图 3.16 所示。

```
$ rosnode list
```

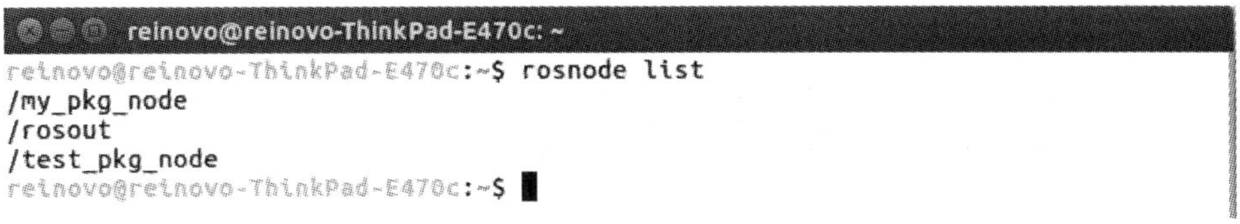

图 3.16 当前系统运行的节点

从运行结果可以看出，系统当前运行了两个节点：my_pkg_node 和 test_pkg_node。

由于在 launch 文件中将 test_pkg_node 赋予了重生属性，输入以下命令模拟关闭 test_pkg_node，然后观察该节点被关闭后是否能重新被打开，如图 3.17 所示。

```
$ rosnode kill /test_pkg_node
```

图 3.17 test_pkg_ndoe 重生结果

3. 必要节点属性(required)

当一个节点被声明为必要节点即 required="true"终止时,roslaunch 会终止所有其他活跃节点并退出。比如,在依赖激光雷达的机器人导航中,若激光雷达节点意外退出时,则 roslaunch 将会终止其他节点然后退出。

修改 startup.launch 文件如下：

```
<launch>
    <node pkg = "test_pkg" type = "test_pkg_node" name = "test_pkg_node" respawn = "true" output = "screen"/>
    <node pkg = "my_pkg" type = "my_pkg_node" name = "my_pkg_node" required = "true" output = "screen"/>
</launch>
```

上述代码中,将 my_pkg_node 节点设置为必要属性,现在关闭 my_pkg_node 节点,如图 3.18 所示。

```
$ rosnode kill /my_pkg_node
```

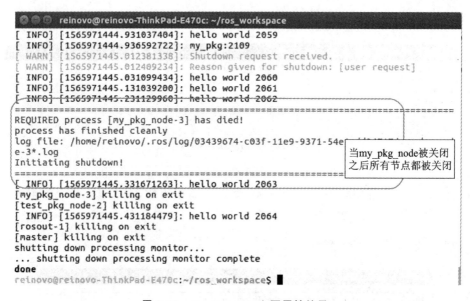

图 3.18 my_pkg_node 必要属性结果

从上面的结果可以看到,必要属性的优先级要高于重生属性。

3.6.3 <include> 标签

在当前 launch 文件中调用另一个 launch 文件,方便代码的复用,可以使用包含(include)标签：

```
<include file = "$(find package-name)/launch-file-name">
```

由于直接输入路径信息很烦琐且容易出错,大多数包含元素都使用查找(find)命令搜索功能包的位置来替代直接输入绝对路径。

为了方便说明,创建一个功能包 third_pkg 并用 robot_bringup 功能包中的 launch 文件来调用该功能包中的 launch 文件。

```
$ cd ~/ros_workspace/src/
$ catkin_create_pkg third_pkg std_msgs roscpp rospy
```

在/third_pkg 路径下创建一个 launch 文件夹:

```
$ cd ~/ros_workspace/src/third_pkg
$ mkdir launch
```

在 third_pkg/src 路径下创建 third_pkg.cpp 源文件,打开 third_pkg.cpp 输入以下测试代码:

```cpp
#include "ros/ros.h"
#include "std_msgs/String.h"
#include <sstream>
int main(int argc,char **argv)
{
ros::init(argc,argv,"third_pkg");
ros::NodeHandle n;// 更新话题的消息格式为自定义的消息格式
ros::Publisher chatter_pub = n.advertise <std_msgs::String>("third_pkg",1000);
ros::Rate loop_rate(10);
int count = 0;
while(ros::ok())
{
std_msgs::String msg;
std::stringstream ss;
ss << "third pkg:" << count;
msg.data = ss.str();
ROS_INFO("%s",msg.data.c_str());
chatter_pub.publish(msg);// 将消息发布到话题中
ros::spinOnce();
loop_rate.sleep();
++count;
}
return 0;
}
```

编辑源码之后同样要修改 CMakeLists.txt,将 third_pkg.cpp 编译为可执行代码,如图 3.19 所示。

编译工作空间。在 third_pkg/launch 路径下新建一个 third_pkg.launch,并输入以下代码:

```xml
<launch>
<node pkg = "third_pkg" type = "third_pkg_node" name = "third_pkg_node" output = "screen"/>
</launch>
```

打开 robot_bringup 功能包下的 startup.launch 文件,修改如下:

```xml
<launch>
<node pkg = "test_pkg" type = "test_pkg_node" name = "test_pkg_node" respawn = "true" output = "screen"/>
<node pkg = "my_pkg" type = "my_pkg_node" name = "my_pkg_node" required = "true" output = "screen"/>
```

智能网联汽车 ROS 实战入门

```
## With catkin_make all packages are built within a single CMake context
## The recommended prefix ensures that target names across packages don't co
llide
add_executable(${PROJECT_NAME}_node src/third_pkg.cpp)

## Rename C++ executable without prefix
## The above recommended prefix causes long target names, the following rena
mes the
## target back to the shorter version for ease of user use
## e.g. "rosrun someones_pkg node" instead of "rosrun someones_pkg someones_
pkg_node"
# set_target_properties(${PROJECT_NAME}_node PROPERTIES OUTPUT_NAME node PRE
FIX "")

## Add cmake target dependencies of the executable
## same as for the library above
# add_dependencies(${PROJECT_NAME}_node ${${PROJECT_NAME}_EXPORTED_TARGETS}
${catkin_EXPORTED_TARGETS})

## Specify libraries to link a library or executable target against
target_link_libraries(${PROJECT_NAME}_node
  ${catkin_LIBRARIES}
)
```

图 3.19 修改 CMakeLists.txt

```
<include file = "$(find third_pkg)/launch/third_pkg.launch"/>
</launch>
```

运行 startup.launch,结果如图 3.20 所示。

```
$ roslaunch robot_bringup startup.launch
```

```
reinovo@reinovo-ThinkPad-E450c:~$ roslaunch robot_bringup startup.launch
... logging to /home/reinovo/.ros/log/1da1a56c-c0b0-11e9-816e-48e24458bd29/rosla
unch-reinovo-ThinkPad-E450c-16566.log
Checking log directory for disk usage. This may take awhile.
Press Ctrl-C to interrupt
Done checking log file disk usage. Usage is <1GB.

started roslaunch server http://reinovo-ThinkPad-E450c:35365/

SUMMARY
========

PARAMETERS
 * /rosdistro: kinetic
 * /rosversion: 1.12.14

NODES
  /
    my_pkg_node (my_pkg/my_pkg_node)        运行了三个节点
    test_pkg_node (test_pkg/test_pkg_node)
    third_pkg_node (third_pkg/third_pkg_node)

auto-starting new master
process[master]: started with pid [16576]
ROS_MASTER_URI=http://localhost:11311

setting /run_id to 1da1a56c-c0b0-11e9-816e-48e24458bd29
process[rosout-1]: started with pid [16589]
started core service [/rosout]
process[test_pkg_node-2]: started with pid [16593]
process[my_pkg_node-3]: started with pid [16602]
process[third_pkg_node-4]: started with pid [16609]
[ INFO] [1566019826.516333738]: test_pkg version:1.0.0
[ INFO] [1566019826.516417989]: hello world 100
[ INFO] [1566019826.517698712]: test_pkg version:1.0.0,init count:100
[ INFO] [1566019826.517786516]: my_pkg:150
```

图 3.20 包含其他 launch 文件运行结果

3.6.4 <param> 标签

在开发程序的过程中,经常会遇到一些随时要调整的参数,而每次在源码中调节参数都需要重新编译源码,这样很麻烦。通过 <param> 标签可以在 launch 文件中直接设定节点中的参数,这样就没有必要修改源码和编译了,每次只要修改一下 launch 文件即可直接修改节点运行的参数。一个 <param> 标签通常写法如下:

```
<param name = "参数的名称" type = "参数的类型" value = "参数的值" />
```

其中:

  name 指定了该参数的名称;
  type 指定了该参数的类型,可选项有:str,int,double,bool,yaml;
  value 参数的具体值。

举例说明,打开 robot_bringup 功能包中的 startup.launch 文件,在 my_pkg 节点中利用 <param> 标签添加一些参数。代码如下:

```xml
<launch>
    <node pkg = "test_pkg" type = "test_pkg_node" name = "test_pkg_node" respawn = "true" output = "screen"/>
    <node pkg = "my_pkg" type = "my_pkg_node" name = "my_pkg_node" required = "true" output = "screen">
        <param name = "name" type = "str" value = "reinovo" />
        <param name = "age" type = "int" value = "13"/>
        <param name = "handsome" type = "bool" value = "true"/>
        <param name = "salary" type = "double" value = "12345678.00"/>
    </node>
    <include file = " $ (find third_pkg)/launch/third_pkg.launch"/>
</launch>
```

在 launch 文件中定义完参数之后,修改 my_pkg.cpp 源码,在源码中接收调用这些参数。

```cpp
#include "ros/ros.h"
#include "std_msgs/String.h"
#include <sstream>
#include "test_pkg/test_pkg.h"
#include "my_pkg/multiply.h" //动态库头文件
using namespace std;
int main(int argc, char **argv)
{
    string my_name = "";
    int my_age = 0;
    bool my_handsome = false;
    double my_salary = 0.0;//定义变量用于接收参数

    ros::init(argc,argv,"my_pkg");
    ros::NodeHandle n;
    ros::param::get("~name",my_name);
    ros::param::get("~age",my_age);
    ros::param::get("~handsome",my_handsome);
    ros::param::get("~salary",my_salary);//获取 launch 文件中的参数
    ros::Publisher chatter_pub = n.advertise <std_msgs::String> ("my_chatter", 1000);
```

```cpp
    ros::Rate loop_rate(10);
    int count = multiply(30,5);  //调用动态库中的方法
    ROS_INFO("Get Param,name:%s,age:%d,isHandsome:%d,salary:%f",my_name.c_str(),my_age,my_handsome,my_salary);
    ROS_INFO("test_pkg version:%s,init count:%d",TEST_PKG_VER,INIT_COUNT);
    while (ros::ok())
    {
        std_msgs::String msg;
        std::stringstream ss;
        ss << "my_pkg:" << count;
        msg.data = ss.str();
        ROS_INFO("%s", msg.data.c_str());
        chatter_pub.publish(msg);
        ros::spinOnce();
        loop_rate.sleep();
        ++count;
    }
    return 0;
}
```

代码中黑色字体为新添加的代码,至于代码中如何获取 launch 文件中的参数,将在第 6 章参数服务器中详细介绍。

修改源码后对工作空间进行编译,编译成功后运行 startup.laucn 文件结果如图 3.21 所示。

图 3.21　参数调用结果

3.6.5 \<arg\> 标签

\<arg\> 标签用于在 launch 文件中定义一些变量以方便修改。通常一个 \<arg\> 标签如下：

```
<arg name = "参数的名字" default = "默认值" doc = "描述"/>
```

举例如下：

```
<launch>
    <arg name = "driver_methods" default = "two_wheels" doc = "driver_type [two_wheels, three_wheels]"/>
    <!-- arg name = "mode" default = "three_wheels"/ -->
    <node pkg = "bobac2_base" type = "bobac2_base_node" name = "bobac2_base_node" output = "screen"/>
    <include file = "$(find bobac2_kinematics)/launch/bobac2_kinematics_$(arg driver_methods).launch"/>
</launch>
```

此例中可以通过参数的配置决定运行哪个 launch 文件。为了更加直观地了解，再举一个例子。打开 robot_bringup 功能包下的 startup.launch，修改如下：

```
<launch>
    <arg name = "my_name" default = "reinovo"/> //添加了一个变量
    <node pkg = "test_pkg" type = "test_pkg_node" name = "test_pkg_node" respawn = "true" output = "screen"/>
    <node pkg = "my_pkg" type = "my_pkg_node" name = "my_pkg_node" required = "true" output = "screen">
        <param name = "name" type = "str" value = "$(arg my_name)"/> //调用变量
        <param name = "age" type = "int" value = "13"/>
        <param name = "handsome" type = "bool" value = "true"/>
        <param name = "salary" type = "double" value = "12345678.00"/>
    </node>
    <include file = "$(find third_pkg)/launch/third_pkg.launch"/>
</launch>
```

在 launch 文件中添加了一个名字为 my_name 的变量，并且把它赋值给了参数 name。运行 launch 文件观察结果。

```
$ roslaunch robot_bringup startup.launch my_name:=qqqq
```

这样即可在运行 launch 文件时，通过定义的变量动态地修改参数的赋值，结果如图 3.22 所示。

3.6.6 \<group\> 标签

\<group\> 标签允许将节点分组放到不同的命名空间，这是一个很实用的标签，比如，要将同一个节点运行两次，就可以将这个节点放到不同的两个 group 中。

```
<group ns = "命名空间的名字" clear_params = "true|false"/>
```

其中：

ns　命名空间的名字；

clear_params　运行前是否清空所有该命名空间下所有的参数。

图 3.22 变量的使用方法

举例说明,打开 robot_bringup 功能包下的 startup.launch 文件修改代码如下:

```
<launch>
<arg name = "my_name" default = "reinovo"/>
<group ns = "test1"> //group1
    <node pkg = "test_pkg" type = "test_pkg_node" name = "test_pkg_node" respawn = "true" output = "screen"/>
    <node pkg = "my_pkg" type = "my_pkg_node" name = "my_pkg_node" required = "true" output = "screen">
        <param name = "name" type = "str" value = " $ (arg my_name)"/>
        <param name = "age"  type = "int" value = "13"/>
        <param name = "handsome" type = "bool" value = "true"/>
        <param name = "salary" type = "double" value = "12345678.00"/>
    </node>
    <include file = " $ (find third_pkg)/launch/third_pkg.launch"/>
</group>
<group ns = "test2"> //group2
    <node pkg = "test_pkg" type = "test_pkg_node" name = "test_pkg_node" respawn = "true" output = "screen"/>
    <node pkg = "my_pkg" type = "my_pkg_node" name = "my_pkg_node" required = "true" output = "screen">
        <param name = "name" type = "str" value = " $ (arg my_name)"/>
        <param name = "age"  type = "int" value = "13"/>
        <param name = "handsome" type = "bool" value = "true"/>
        <param name = "salary" type = "double" value = "12345678.00"/>
    </node>
```

```
<include file = " $ (find third_pkg)/launch/third_pkg.launch"/>
</group>
</launch>
```

运行 launch 文件查看 <group> 运行结果，如图 3.23 所示。

```
$ roslaunch robot_bringup startup.launch
```

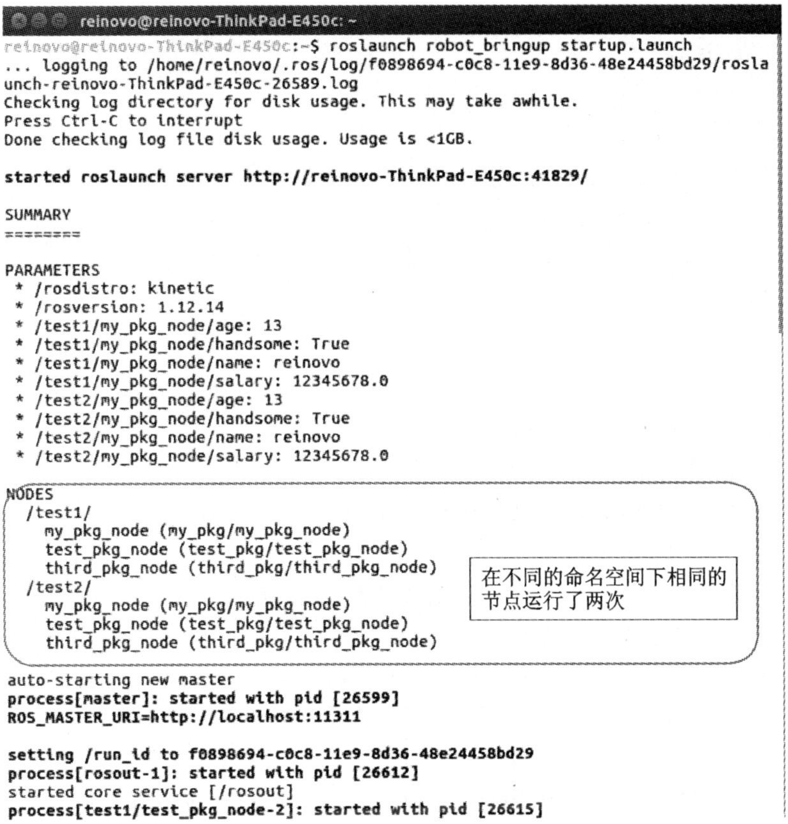

图 3.23 <group> 运行结果

查看当前系统中运行的节点。图 3.24 中可以看到运行的节点被批量重新命名，前面加上了命名空间的名字。

```
$ rosnode list
```

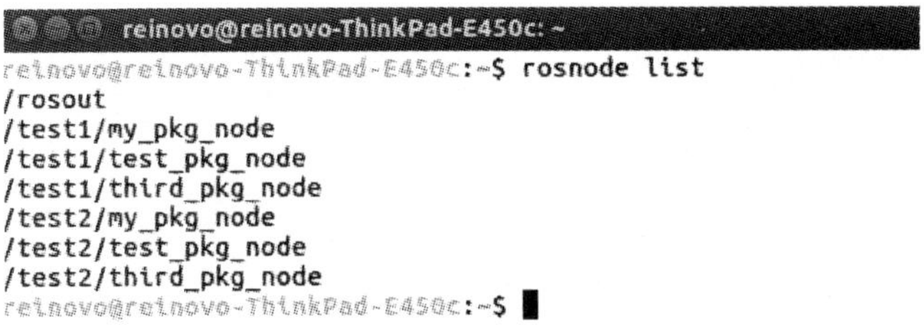

图 3.24 节点被重新命名

课后练习

选择题

(1) [单选]在 CMake 文件编写规则中,用于将库文件链接到目标文件的指令是(　　)。
　　(A) add_executable()
　　(B) add_link()
　　(C) add_library()
　　(D) target_link_libraries()

(2) [单选]在 CMake 的指令中,引入头文件搜索路径的指令是(　　)。
　　(A) cmake_include_directory()
　　(B) cmake_include_path()
　　(C) include_directories()
　　(D) include_directory()

(3) [单选]关于 ROS Node 的描述,错误的是(　　)。
　　(A) Node 启时会向 Master 注册
　　(B) Node 可以先于 ROS Master 启动
　　(C) Node 是 ROS 可执行文件运行的实例
　　(D) Node 是 ROS 的进程

(4) [单选]关于.launch 文件的描述,错误的是(　　)。
　　(A) 可以加载配置好的参数,方便快捷
　　(B) 通过 roslaunch 命令来启动 launch 文件
　　(C) 在 roslaunch 前必须先 roscore
　　(D) 可以一次性启动多个节点,减少操作

(5) [单选]如果你要复制一个 ROS 的软件包,下列路径存放位置合理的是(　　)。
　　(A) ~/catkin_ws/
　　(B) ~/catkin_ws/devel
　　(C) ~/catkin_ws/build
　　(D) ~/my_ws/src

(6) [多选]一个 ROS 的 pacakge 要正常的编译,必须要有的文件是(　　)。
　　(A) *.cpp
　　(B) CMakeLists.txt
　　(C) *.h
　　(D) package.xml

第 4 章
ROS 通信机制——话题

4.1 话题通信原理

在 ROS 的通信方式中，话题是常用的一种。对于实时性、周期性的消息，使用话题来传输是最佳的选择。话题是一种点对点的单向通信方式，这里的"点"指的是节点，也就是说节点之间可以通过话题方式来传递信息。话题要经历下面几步的初始化过程：首先，发布者节点和订阅者节点都要到节点管理器进行注册；然后，发布者会发布话题，订阅者在节点管理器的指挥下会订阅该话题，从而建立起订阅-发布之间的通信。注意整个过程是单向的。其结构示意图如图 4.1 所示。

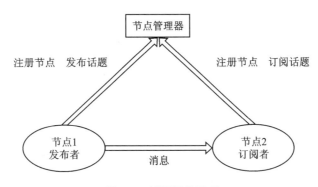

图 4.1 话题通信原理

订阅者接收消息会进行处理，一般这个过程叫做回调（Callback）。所谓回调就是提前定义好一个处理函数（写在代码中），当有消息来时就会触发这个处理函数，函数会对消息进行处理。

图 4.1 就是 ROS 的话题通信方式的流程示意图。话题通信属于一种异步的通信方式。下面通过一个示例来了解如何使用话题通信。

如图 4.2 所示，我们以摄像头画面的发布、处理、显示为例介绍话题通信的流程。在机器人上的摄像头拍摄程序是一个节点（圆圈表示，记作节点 1），当节点 1 运行启动之后，它作为一个发布者就开始发布话题。比如它发布了一个话题（方框表示）叫做/camera_rgb，是 rgb 颜色信息，即采集到的彩色图像。同时，节点 2 假如是图像处理程序，它订阅了/camera_rgb 这个话题，经过节点管理器的介绍，它就能建立与摄像头节点（节点 1）的连接。

那么，"异步"这个概念怎么理解呢？在节点 1 每发布一次消息之后，就会继续执行下一个动作，至于消息是什么状态、被怎样处理，它不需要了解；而对于节点 2 图像处理程序，它只负责接收和处理/camera_rgb 上的消息，至于是谁发来的，它不会关心。所以，节点 1、节点 2 两者都是各司其职，不存在协同工作，我们称这样的通信方式是异步的。

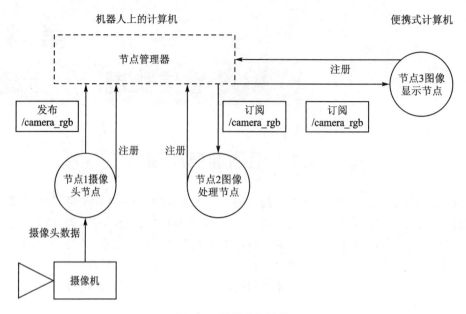

图 4.2 话题通信流程

ROS 是一种分布式的架构,一个话题可以被多个节点同时发布,也可以同时被多个节点接收。比如在这个场景中用户可以再加入一个图像显示的节点,如果再想看看摄像头节点的画面,则可以将自己的便携式计算机连接到机器人上的节点管理器,然后在自己的计算机上启动图像显示节点。

这就体现了分布式系统通信的好处:扩展性好、软件复用率高。

总结三点:

① 话题通信方式是异步的,发送时调用 publish()方法,发送完成立即返回,不用等待反馈。

② 订阅者通过回调函数的方式来处理消息。

③ 话题可以同时有多个订阅者,也可以同时有多个发布者。

4.2 话题通信示例

继续前面的示例来深入理解一下什么是话题。之前教程中 ros_workspace 中创建的三个软件包都是发布者,我们需要酌情修改一下,然后增加一个订阅者节点,这样就能真实地感受到 ROS 中节点是如何通过话题进行通信了。

输入以下命令,在 ros_workspace 中创建第四个功能包 subscribe_pkg:

```
$ cd ~/ros_workspace/src/
$ catkin_create_pkg subscribe_pkg rospy roscpp std_msgs
```

接下来修改一下 third_pkg 软件包的源码 third_pkg.cpp,由于最初的代码编写时发送的话题类型为字符串,这里需要将其修改为整型,这样方便后面使用 rqt_plot 画图软件将话题数据可视化演示,修改如下:

① 将默认的话题类型从 String 修改为 Int32;

② 将话题名称从 third_pkg 修改为 third_pkg_topic,方便与节点 third_pkg 区别开;
③ 将发送话题的频率从 10 Hz 改为 2 Hz,500 ms 发送一次;
④ 话题中的数据内容从 0 到 4 循环发送,方便查看动态发布过程。

修改后的 third_pkg.cpp 代码如下:

```cpp
#include "ros/ros.h"
#include "std_msgs/Int32.h"
int main(int argc,char **argv)
{
ros::init(argc,argv,"third_pkg");
ros::NodeHandle n;// 更新话题的消息格式为自定义的消息格式
ros::Publisher chatter_pub = n.advertise <std_msgs::Int32>("third_pkg_topic",1000);
ros::Rate loop_rate(2);
int count = 0;
while(ros::ok())
{
std_msgs::Int32 msg;
msg.data = count;
ROS_INFO("%d",msg.data);
chatter_pub.publish(msg);//将消息发布到话题中
ros::spinOnce();
loop_rate.sleep();
count = (++count)%5;
}
return 0;
}
```

在功能包 subscribe_pkg/src 路径下创建 subscribe.cpp,并写入以下代码:

```cpp
#include "ros/ros.h"
#include "std_msgs/Int32.h"

void chatterCallback(const std_msgs::Int32::ConstPtr& msg)
 {
ROS_INFO("I heard name:[%d]",msg-> data);
}
int main(int argc,char **argv)
{
  ros::init(argc,argv,"subscribe_node");
  ros::NodeHandle n;
  ros::Subscriber sub = n.subscribe("third_pkg_topic",1000,chatterCallback);
  ros::spin();
  return 0;
}
```

修改 subscribe_pkg 下的 CMakeLists.txt 文件,如图 4.3 所示。
编译工作空间:

```
$ cd ~/ros_workspace
$ catkin_make
```

修改 third_pkg 功能包下的 third_pkg.launch 文件,如图 4.4 所示。
运行 third_pkg.launch,结果如图 4.5 所示。

```
$ roslaunch third_pkg third_pkg. launch
```

```
reinovo@reinovo-ThinkPad-E450c: ~/ros_workspace
## With catkin_make all packages are built within a single CMake context
## The recommended prefix ensures that target names across packages don't co
llide
 add_executable(${PROJECT_NAME}_node src/subscribe.cpp)

## Rename C++ executable without prefix
## The above recommended prefix causes long target names, the following rena
mes the
## target back to the shorter version for ease of user use
## e.g. "rosrun someones_pkg node" instead of "rosrun someones_pkg someones_
pkg_node"
# set_target_properties(${PROJECT_NAME}_node PROPERTIES OUTPUT_NAME node PRE
FIX "")

## Add cmake target dependencies of the executable
## same as for the library above
# add_dependencies(${PROJECT_NAME}_node ${${PROJECT_NAME}_EXPORTED_TARGETS}
${catkin_EXPORTED_TARGETS})

## Specify libraries to link a library or executable target against
 target_link_libraries(${PROJECT_NAME}_node
   ${catkin_LIBRARIES}
 )
:set nu                                                   151,1          73%
```

图 4.3　subscribe_pkg 下的 CMakeLists. txt 文件

```
reinovo@reinovo-ThinkPad-E450c: ~/ros_workspace
<launch>
<node pkg="third_pkg" type="third_pkg_node" name="third_pkg_node" output="screen
"/>
<node pkg="subscribe_pkg" type="subscribe_pkg_node" name="subscribe_pkg" output=
"screen"/>
</launch>
~
~
                                                          1,1            All
```

图 4.4　third_pkg. launch 文件修改

```
reinovo@reinovo-ThinkPad-E450c: ~/ros_workspace
setting /run_id to 0862ecc8-c31d-11e9-8a19-48e24458bd29
process[rosout-1]: started with pid [24051]
started core service [/rosout]
process[third_pkg_node-2]: started with pid [24054]
process[subscribe_pkg-3]: started with pid [24062]
[ INFO] [1566286508.201456057]: 0
[ INFO] [1566286508.701304748]: 1
[ INFO] [1566286508.701884935]: I heard name:[1]
[ INFO] [1566286509.201293647]: 2
[ INFO] [1566286509.201741391]: I heard name:[2]
[ INFO] [1566286509.701291168]: 3
[ INFO] [1566286509.701665658]: I heard name:[3]
[ INFO] [1566286510.201266033]: 4
[ INFO] [1566286510.201566232]: I heard name:[4]
[ INFO] [1566286510.701312005]: 0
[ INFO] [1566286510.701787614]: I heard name:[0]
[ INFO] [1566286511.201233309]: 1
[ INFO] [1566286511.201687093]: I heard name:[1]
[ INFO] [1566286511.701313161]: 2
[ INFO] [1566286511.701880188]: I heard name:[2]
[ INFO] [1566286512.201297854]: 3
[ INFO] [1566286512.201678158]: I heard name:[3]
^C[ INFO] [1566286512.701211937]: 4
[ INFO] [1566286512.701420202]: I heard name:[4]
[subscribe_pkg-3] killing on exit
[third_pkg_node-2] killing on exit
```

图 4.5　third_pkg. launch 运行结果

该示例由 third_pkg_node 节点发布话题 third_pkg_topic，subscribe_pkg 订阅该话题，并将该话题传输的信息打印到终端。

4.3 rqt_graph 和 rqt_plot 命令的使用

rqt_graph 工具可以用来显示当前系统中运行的节点以及各节点之间通信的话题。

打开一个终端输入以下命令：

```
$ rqt_graph
```

图 4.6 中椭圆代表运行的节点，长方形代表话题，箭头表示话题传输的方向。当我们拿到一个陌生的机器人工程时，可以用 rqt_graph 快速地查看系统当前的信号流图，了解系统各节点的输入/输出情况，有助于我们快速地了解系统。

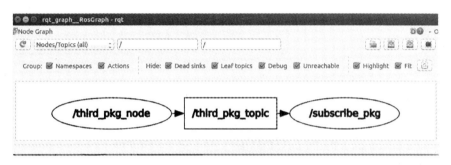

图 4.6 节点通信图

rqt_plot 命令允许我们将话题的值以二位曲线的形式展示，相当于一个数字示波器，辅助我们开发机器人。运行结果如图 4.7 所示。

```
$ rqt_plot
```

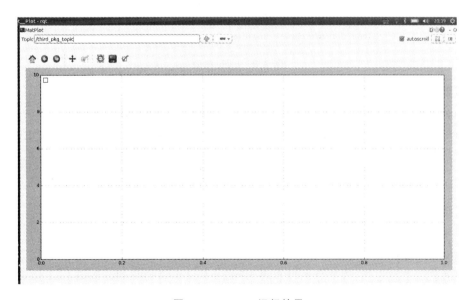

图 4.7 rqt_plot 运行结果

在 Topic 话题栏中写入要展示的话题，单击"＋"即可以图形显示要展示的话题，如图 4.8 所示。

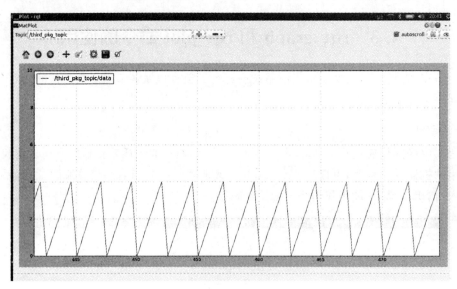

图 4.8　rqt_plot 展示话题/third_pkg_topic

4.4　rostopic 命令的使用

ROS 同时提供了一系列命令方便开发者在终端操作话题，如图 4.9 所示。

```
Commands:
    rostopic bw      display bandwidth used by topic
    rostopic delay   display delay of topic from timestamp in header
    rostopic echo    print messages to screen
    rostopic find    find topics by type
    rostopic hz      display publishing rate of topic
    rostopic info    print information about active topic
    rostopic list    list active topics
    rostopic pub     publish data to topic
    rostopic type    print topic or field type
```

图 4.9　topic 命令

在这里介绍几个比较常用的命令。

1. rostopic list

此命令可以打印出当前系统中存在的所有话题名称。假设上面例子的 third_pkg.launch 还在运行，打开一个终端，输入以下命令：

```
$ rostopic list
```

运行结果如图 4.10 所示。

从图 4.10 中可以看出，当前系统一共有三个话题。

2. rostopic info

查看消息的信息包括：该消息的发布者、订阅者及该消息的类型。例如：

```
$ rostopic info /third_pkg_topic
```

```
reinovo@reinovo-ThinkPad-E450c: ~
reinovo@reinovo-ThinkPad-E450c:~$ rostopic list
/rosout
/rosout_agg
/third_pkg_topic
reinovo@reinovo-ThinkPad-E450c:~$
```

图 4.10　rostopic list 命令结果

示例结果如图 4.11 所示。

```
reinovo@reinovo-ThinkPad-E450c: ~
reinovo@reinovo-ThinkPad-E450c:~$ rostopic info /third_pkg_topic
Type: std_msgs/Int32

Publishers:
 * /third_pkg_node (http://reinovo-ThinkPad-E450c:35357/)

Subscribers:
 * /subscribe_pkg (http://reinovo-ThinkPad-E450c:40119/)


reinovo@reinovo-ThinkPad-E450c:~$
```

图 4.11　rostopic info 示例结果

3．rostopic echo

该命令可以打印指定话题的内容到终端，例如：

```
$ rostopic echo /third_pkg_topic
```

示例结果如图 4.12 所示。

```
reinovo@reinovo-ThinkPad-E450c: ~
reinovo@reinovo-ThinkPad-E450c:~$ rostopic echo /third_pkg_topic
data: 2
---
data: 3
---
data: 4
---
data: 0
---
data: 1
---
data: 2
---
data: 3
---
data: 4
```

图 4.12　rostopic echo 示例结果

4．rostopic pub

该命令可以通过终端向话题发布消息，例如：

```
$ rostopic pub -r 10 /third_pkg_topic std_msgs/Int32 "data: 100"
```

其中：

-r 10 指定发布的频率为 10；

std_msgs/Int32 "data: 100"为发布消息的内容。

注意：这里在发布一个话题时可以利用 Tab 键来补全该命令，比如：$ rostopic pub - r 10 /third_pkg_topic <此时可以按 Tab 键> 即可补全后面的内容。

此时再利用 rostopic echo /third_pkg_topic 查看发布内容，如图 4.13 所示。

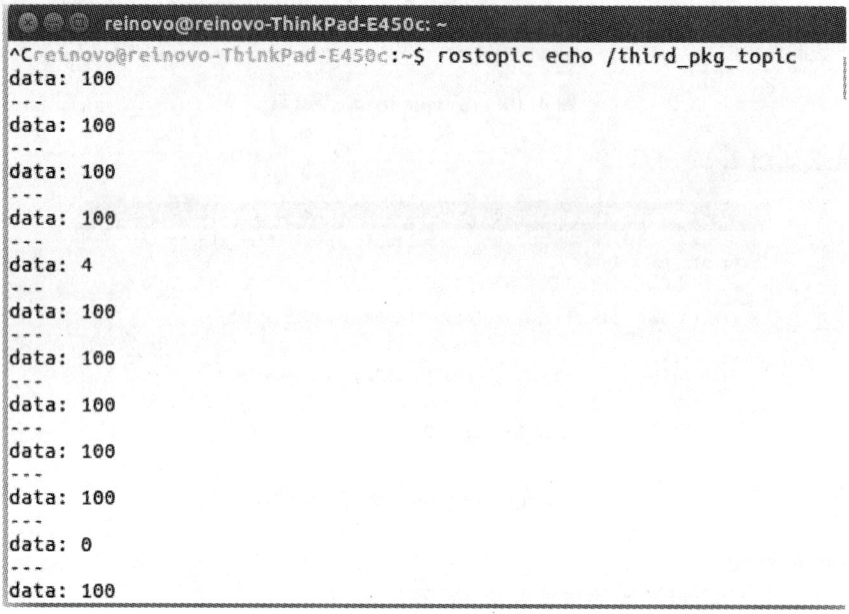

图 4.13 rostopic echo 命令结果

图 4.13 中打印了当前 third_pkg_topic 话题上传递的消息。

5. rostopic type

该命令可以输出指定话题的类型，即该话题上传输的信息的数据类型。图 4.14 中所示的命令 std_msgs/Int32 表明/third_pkg_topoc 话题上只能传输 int32 的数据。

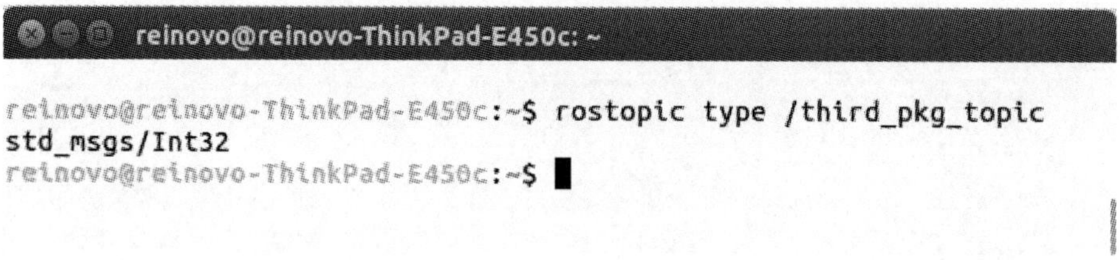

图 4.14 rostopic type /third_pkg_topic 命令

4.5 代码解析

4.5.1 third_pkg.cpp 源码分析

third_pkg.cpp 源码分析如图 4.15 所示。

头文件 ros/ros.h 是一个实用的头文件，它引用了 ROS 系统中大部分常用的头文件。头文件 std_msgs/Int32.h 包含 msgs/Int32 消息，是系统提供的标准消息类型。

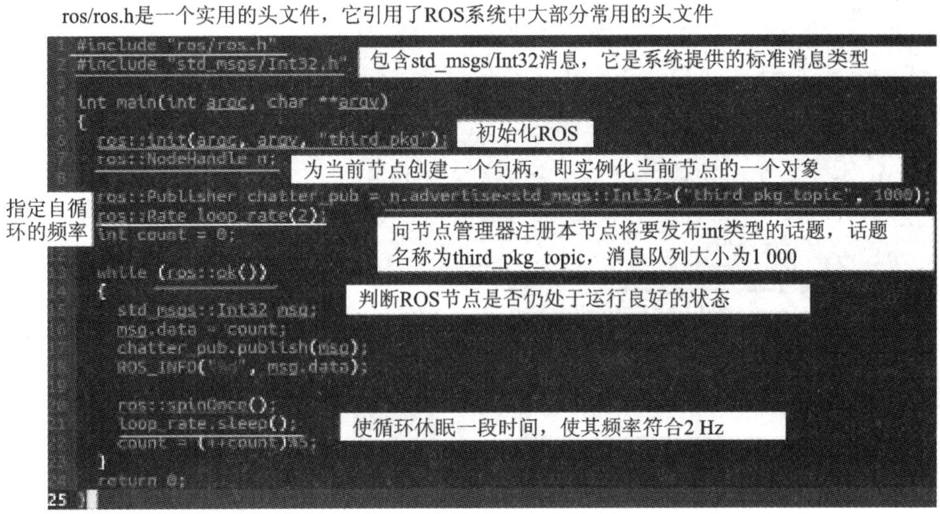

图 4.15 third_pkg.cpp 源码分析

ros::init()函数为初始化 third_pkg 节点。

ros::NodeHandle n 为 third_pkg 节点创建一个句柄,为实例化 third_pkg 节点的一个对象 n。

ros::Publisher chatter_pub=n.advertise<std_msgs::String>("third_pkg",1000)表示发布话题的名称是 third_pkg,消息队列的最大长度为 1 000,如果发布的消息超过这个长度而没有被接收,那么旧的消息就会出队。

ros::Rate loop_rate(2)表示发送话题的频率为 2 Hz,即 500 ms 发送一次。

while(ros::ok())表示判断 ROS 节点是否处于运行良好的状态。

std_msgs::Int32 msg 表示话题类型为 Int32。

ROS_INFO("%d",msg.data)表示将 msg.data 数据信息以整型格式打印出来。

loop_rate.sleep()表示循环休眠 500 ms。

4.5.2 subscrbe.cpp 源码分析

subscrbe.cpp 源码分析如图 4.16 所示。

chatterCallback()作为回调函数相当于一个"中断"子程序,当订阅的话题中有收到消息时自动调用,消息是以指针的形式传输。

n.subscribe("third_pkg_topic",1000,chatterCallback)函数用于通知主节点的节点管理器在话题 third_pkg_topic 上订阅消息,节点管理器是一个主管发布和订阅的记录。通知节点管理器之后,节点管理器会给出响应,相当于通知可以在这个话题 third_pkg_topic 上订阅消息了,subscribe 的返回值返回一个 subscriber 类型的对象 sub,到此为止这个节点才真正有了订阅者的功能。功能的实现依赖于 subscriber 类中包含的成员函数。subscriber 采用一种类似于中断的模式,使用 ros::spin()函数进入自循环,等待消息到来后就去调用消息回调函数 chatterCallback()。

图 4.16 subscrbe.cpp 源码分析

4.6 理解自定义消息类型

4.6.1 什么是消息

在前面几节中我们了解了什么是话题,话题传输的原理,学习了如何编写一个话题的发布者和订阅者。话题是 ROS 提供的一种重要通信方式,话题上传输的信息内容即为消息。话题好比一个车道,消息就是跑在车道上的汽车。只不过 ROS 规定了一个"车道"上只能跑一种"车型"。

话题的类型由消息的数据类型决定,ROS 提供了组成消息的基本数据类型:
- int8,int16,int32,int64(plus uint *);
- float32,float64;
- string;
- time,duration;
- other msg files;
- variable-length array[],fixed-length array[C]。

4.6.2 rosmsg 命令

ROS 提供了一系列命令如图 4.17 所示,帮助我们来操作消息。

1. rosmsg list

列出当前 ROS 系统中所有的消息格式,通过 wc 可以统计共有多少条消息。

rosmsg list | wc -l:统计当前系统共有多少条命令。

2. rosmsg show

该命令可以展示消息的定义格式。

```
reinovo@reinovo-ThinkPad-E450c: ~
reinovo@reinovo-ThinkPad-E450c:~$ rosmsg -h
rosmsg is a command-line tool for displaying information about ROS Message types
.

Commands:
        rosmsg show     Show message description
        rosmsg info     Alias for rosmsg show
        rosmsg list     List all messages
        rosmsg md5      Display message md5sum
        rosmsg package  List messages in a package
        rosmsg packages List packages that contain messages

Type rosmsg <command> -h for more detailed usage

reinovo@reinovo-ThinkPad-E450c:~$ 
```

图 4.17　rosmsg 系列命令

如图 4.18 所示，std_msgs/Int32 消息的格式为一个 int32 的数据；geometry_msgs/Twist 命令包含 linear 和 angular 两个复合域。一个复合域是由简单的一个或多个子域组合而成，其中的每一个子域可能是另一个复合域或者独立域，而且它们也都由基本数据类型组成。

```
reinovo@reinovo-ThinkPad-E450c:~$ rosmsg show std_msgs/Int32
int32 data

reinovo@reinovo-ThinkPad-E450c:~$ rosmsg show geometry_msgs/Twist
geometry_msgs/Vector3 linear
  float64 x
  float64 y
  float64 z
geometry_msgs/Vector3 angular
  float64 x
  float64 y
  float64 z
```

图 4.18　rosmsg show 命令

4.7　创建自定义消息类型

一个消息的定义类似于 C 语言的结构体。ROS 允许我们自定义消息以满足通过话题传输想要的数据。下面通过示例来介绍如何自定义一个消息。

① 在 third_pkg 功能包下创建 msg 文件，并在 msg 文件夹下新建 myTestMsg.msg 文件。创建完成后的 third_pkg 功能包结构如图 4.19 所示。

```
reinovo@reinovo-ThinkPad-E450c:~/ros_workspace/src$ tree third_pkg/
third_pkg/
├── CMakeLists.txt
├── include
│   └── third_pkg
├── launch
│   └── third_pkg.launch
├── msg
│   └── myTestMsg.msg
├── package.xml
└── src
    └── third_pkg.cpp

5 directories, 5 files
reinovo@reinovo-ThinkPad-E450c:~/ros_workspace/src$ 
```

图 4.19　third_pkg 功能包结构

② 打开 myTestMsg.msg 文件输入如图 4.20 所示的内容。

```
reinovo@reinovo-ThinkPad-E450c: ~/ros_workspace/src/third_pkg/msg
string name
int32 age
bool handsome
float32 salary
~
~
~
```

图 4.20　myTestMsg.msg 文件内容

③ 修改 CMakeLists.txt 以及 package.xml 文件。

CMakeLists.txt 文件中需要修改三个地方，首先添加对 message_generation 的依赖，其次添加自定义消息文件，最后添加消息生成依赖。具体修改内容如图 4.21 所示。

```
## Find catkin macros and libraries
## if COMPONENTS list like find_package(catkin REQUIRED COMPONENTS xyz)
## is used, also find other catkin packages
find_package(catkin REQUIRED COMPONENTS
  roscpp
  rospy
  std_msgs
  message_generation
)


## Generate messages in the 'msg' folder
 add_message_files(
   FILES
   myTestMsg.msg
 )


## Generate added messages and services with any dependencies listed here
 generate_messages(
   DEPENDENCIES
   std_msgs  # Or other packages containing msgs
)
```

图 4.21　CMakeLists.txt 文件修改内容

package.xml 文件中只需要添加 message_generation 的依赖即可，如图 4.22 所示加下画线标记部分。

```
<buildtool_depend>catkin</buildtool_depend>
<build_depend>roscpp</build_depend>
<build_depend>rospy</build_depend>
<build_depend>std_msgs</build_depend>
<build_depend>message_generation</build_depend>
<build_export_depend>roscpp</build_export_depend>
<build_export_depend>rospy</build_export_depend>
<build_export_depend>std_msgs</build_export_depend>
<build_export_depend>message_generaion</build_export_depend>
<exec_depend>roscpp</exec_depend>
<exec_depend>rospy</exec_depend>
<exec_depend>std_msgs</exec_depend>
<exec_depend>message_generation</exec_depend>
```

图 4.22　package.xml 文件修改内容

④ 编译工作空间：

```
$ cd ~/ros_workspace
$ catkin_make
```

⑤ 查看生成的自定义消息，图 4.23 所示系统已经识别到我们自定义的消息。

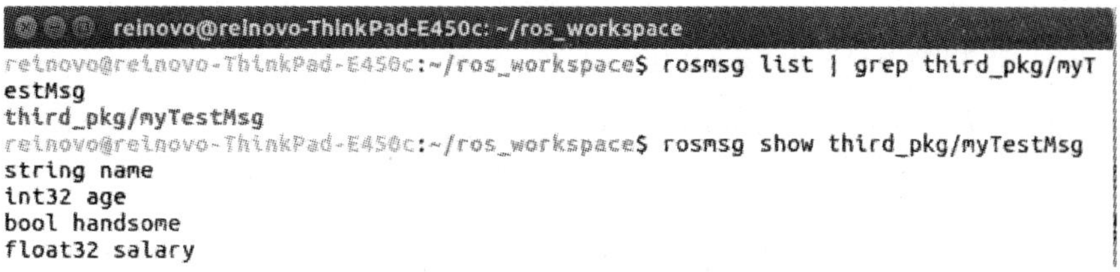
reinovo@reinovo-ThinkPad-E450c: ~/ros_workspace
reinovo@reinovo-ThinkPad-E450c:~/ros_workspace$ rosmsg list | grep third_pkg/myT
estMsg
third_pkg/myTestMsg
reinovo@reinovo-ThinkPad-E450c:~/ros_workspace$ rosmsg show third_pkg/myTestMsg
string name
int32 age
bool handsome
float32 salary

图 4.23 查看自定义消息

同时将在 ros_workspace/devel/include 路径下生成 third_pkg/myTestMsg.h。

⑥ 修改 third_pkg.cpp，调用测试上面生成的自定义消息类型，修改 third_pkg.cpp 代码如下（黑色字体部分是修改部分，要格外注意）：

```
#include "ros/ros.h"
#include "third_pkg/myTestMsg.h"//添加自定义消息的头文件
int main(int argc,char **argv)
{
ros::init(argc,argv,"third_pkg");
ros::NodeHandle n;// 更新话题的消息格式为自定义的消息格式
ros::Publisher chatter_pub = n.advertise <third_pkg::myTestMsg> ("third_pkg_topic",1000);
ros::Rate loop_rate(2);
while(ros::ok())
{
third_pkg::myTestMsg msg;//声明一个自定义消息的对象
msg.name = "corvin";//填充消息内容
msg.age = 20;
msg.handsome = true;
msg.salary = 123.45;
chatter_pub.publish(msg);//将消息发布到话题中
ros::spinOnce();
loop_rate.sleep();
}
return 0;
}
```

修改订阅端代码 subscribe.cpp 如下（黑色字体为修改部分，要格外注意）：

```
#include"ros/ros.h"
#include"third_pkg/myTestMsg.h"
void chatterCallback(const third_pkg::myTestMsg ::ConstPtr& msg)
{
ROS_INFO("I heard - name: %s,age: %d,ishandsome: %d,salary: %f",msg -> name.c_str(),msg ->
age,msg -> handsome,msg -> salary);
}
int main(int argc, char **argv)
{
ros::init(argc,argv,"subscribe_node");
ros::NodeHandle n;
ros::Subscriber sub = n.subscribe("third_pkg_topic",1000,chatterCallback);
ros::spin();
return 0;
}
```

⑦ 添加 subscribe_pkg 对于 third_pkg 的依赖。因为 subscribe.cpp 中引用了 third_pkg/myTestMsg.h 头文件,因此 subscribe_pkg 需要依赖于 third_pkg。分别修改 subscribe_pkg 的 CMakeLists.txt 以及 package.xml,如图 4.24 所示。

```
51  <buildtool_depend>catkin</buildtool_depend>
52  <build_depend>roscpp</build_depend>
53  <build_depend>rospy</build_depend>
54  <build_depend>std_msgs</build_depend>
55  <build_depend>third_pkg</build_depend>
56  <build_export_depend>roscpp</build_export_depend>
57  <build_export_depend>rospy</build_export_depend>
58  <build_export_depend>std_msgs</build_export_depend>
59  <build_export_depend>third_pkg</build_export_depend>
60  <exec_depend>roscpp</exec_depend>
61  <exec_depend>rospy</exec_depend>
62  <exec_depend>std_msgs</exec_depend>
63  <exec_depend>third_pkg</exec_depend>
64
```
package.xml 文件的修改

```
10 find_package(catkin REQUIRED COMPONENTS
11    roscpp
12    rospy
13    std_msgs
14    third_pkg
15 )
```
CMakelists.txt 文件的修改

图 4.24 文件修改

⑧ 编译和运行测试代码,编译工作空间,并运行 third_pkg.launch。

```
$ cd ~/ros_workspace
$ catkin_make
$ roslaunch third_pkg third_pkg.launch
```

运行结果如图 4.25 所示。

```
process[subscribe_pkg-3]: started with pid [24764]
[ INFO] [1566731193.732867497]: I heard-name:corvin,age:20,ishandsome:1,salary:1
23.449997
[ INFO] [1566731194.232805009]: I heard-name:corvin,age:20,ishandsome:1,salary:1
23.449997
[ INFO] [1566731194.732830276]: I heard-name:corvin,age:20,ishandsome:1,salary:1
23.449997
[ INFO] [1566731195.232721457]: I heard-name:corvin,age:20,ishandsome:1,salary:1
23.449997
[ INFO] [1566731195.732826893]: I heard-name:corvin,age:20,ishandsome:1,salary:1
23.449997
[ INFO] [1566731196.232832461]: I heard-name:corvin,age:20,ishandsome:1,salary:1
23.449997
[ INFO] [1566731196.732835273]: I heard-name:corvin,age:20,ishandsome:1,salary:1
23.449997
```

图 4.25 运行结果

课后练习

一、选择题

(1) [单选]想要查看/odom 话题发布的内容,应该用的命令是(   )。

(A) rostopic echo /odom

(B) rostopic content /odom

(C) rostopic info /odom

　　　　(D) rostopic print /odom

(2)［单选］请练习 rosmsg 命令,下列不属于 std_msgs 下的消息是(　　)。

　　　　(A) std_msgs/LaserScan

　　　　(B) std_msgs/Header

　　　　(C) std_msgs/Time

　　　　(D) std_msgs/Float32

(3)［多选］下列哪些是 CMake 没有,而 Catkin 有(Catkin 扩展了)的指令?(　　)

　　　　(A) add_action_files()

　　　　(B) add_message_files()

　　　　(C) add_service_files()

　　　　(D) generate_messages()

(4)［多选］关于话题通信的描述,正确的选项有(　　)。

　　　　(A) 话题是一种异步通信机制

　　　　(B) 一个话题至少要有一个发布者和一个接收者

　　　　(C) 查看当前活跃的话题可以通过 rostopic list 命令

　　　　(D) 一个节点最多只能发布一个话题

二、判断题

同一个话题上可以有多个发布者(　　)。

(A) 正确

(B) 错误

三、操作题

编写一个 ROS 节点利用手柄控制小乌龟运动,实现以下功能:

(1) 借助 ROS 功能包 joy 获取左右摇杆手柄键值;

(2) 左摇杆上下推动发布 x 方向线速度指令,右摇杆左右推动发布绕 z 轴角速度指令;

(3) 控制小乌龟移动。

第 5 章
ROS 通信机制——服务

5.1 认识服务基本概念

在第 4 章中介绍了 ROS 通信方式中的话题通信,我们知道话题是 ROS 中的一种单向的异步通信方式。然而,有些时候单向的通信满足不了通信要求,比如当一些节点只是临时而非周期性的需要某些数据,如果用话题通信方式就会消耗大量不必要的系统资源,造成系统的低效率高功耗。在这种情况下,就需要有另外一种请求-查询式的通信模型。本节介绍 ROS 通信中的另一种通信方式——服务(service)。

为了解决以上问题,服务方式在通信模型上与话题做了区别。服务通信是双向的,它不仅可以发送消息,同时还会有反馈。所以服务包括两部分:一部分是请求方(clinet),另一部分是应答方/服务提供方(server)。这时请求方就会发送一个请求,等待应答方处理,应答方反馈回一个应答,这样通过类似"请求-应答"的机制完成整个服务通信。

这种服务通信方式的示意图如图 5.1 所示。

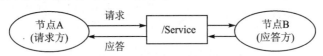

图 5.1 服务通信方式

节点 B 是应答方,提供了一个服务的接口,叫做/Service,一般都会用 string 类型来指定服务的名称,类似于话题。节点 A 向节点 B 发起了请求,经过处理后得到了反馈。

服务是同步通信方式。所谓同步就是说,此时节点 A 发布请求后会在原地等待应答,直到节点 B 处理完请求并且完成了应答,节点 A 才会继续执行。在节点 A 等待过程中,是处于阻塞状态的通信。这样的通信模型没有频繁的消息传递,没有冲突与高系统资源的占用,只有接受请求才执行服务,简单而且高效。

下面对比一下这两种最常用的通信方式,加深对两者的理解和认识,具体如表 5.1 所列。

表 5.1 话题与服务的对比

名 称	话 题	服 务
通信方式	异步通信	同步通信
实现原理	TCP/IP	TCP/IP
通信模型	Publish – Subscribe	Request – Reply
映射关系	Publish – Subscribe(多对多)	Request – Reply(多对一)
特点	接收者收到数据会回调(Callback)	远程过程调用(RPC)服务器端的服务
应用场景	连续、高频的数据发布	偶尔使用的功能/具体的任务
举例	激光雷达、里程计发布数据	开关传感器、拍照、逆解计算

注意：远程过程调用（Remote Procedure Call，RPC），可以简单地理解为在一个进程中调用另一个进程的函数。

5.2 编写 ROS 服务示例

为了更好地理解服务，下面编写一个例子，本例程模拟查询机器人电池的电量。

1. 创建 myTestSrv.srv 文件

在 third_pkg 功能包下创建 srv 工作目录，并在该目录下创建 myTestSrv.srv 文件，创建完成之后，third_pkg 功能包的结构如图 5.2 所示。

图 5.2　third_pkg 功能包结构

打开 myTestSrv.srv 文件输入以下内容：

int32 index
---
int32 result

myTestSrv.srv 文件中包含请求和响应两个部分，用"---"分开。上面部分定义了请求部分的参数，在本例中表示要查询的电池的编号；下面部分表示响应部分，在这里表示返回的电池的电量。**注意**，请求和响应部分的顺序不能写反。

2. 修改 third_pkg 功能包的 CMakeLists.txt

这里与自定义消息的修改方法类似：

① 添加 message_generation 依赖。

```
## is used, also find other catkin packages
find_package(catkin REQUIRED COMPONENTS
  roscpp
  rospy
  std_msgs
  message_generation
  geometry_msgs
)
```

② 添加 .srv 服务文件。

```
## Generate services in the 'srv' folder
  add_service_files(
    FILES
```

```
      myTestSrv.srv
#     Service2.srv
)
```

③ 添加消息生成依赖。

```
## Generate added messages and services with any dependencies listed here
  generate_messages(
    DEPENDENCIES
    std_msgs
    geometry_msgs
)
```

注意：前面创建自定义消息时已经修改过第①步和第③步，这里不需要重复修改。

3. 编译工作空间，生成服务头文件

```
$ cd ~/ros_workspace
$ catkin_make
```

正常情况下将在~/ros_workspace/devel/include/third_pkg 文件夹下生成服务头文件，如图 5.3 所示。

图 5.3 编译生成的服务头文件

4. 修改代码

修改 third_pkg.cpp 代码如下：

```
#include "ros/ros.h"
#include "third_pkg/myTestMsg.h"    //添加自定义消息的头文件
#include "third_pkg/myTestSrv.h"    //添加自定义服务的头文件
static int battery = 100;
bool checkBattery(third_pkg::myTestSrv::Request &req,third_pkg::myTestSrv::Response &res)
                                                                                //服务回调函数
{
if(1 == req.index)
{
ROS_WARN("service sending response:[%d]",battery);
res.result = battery;
}
return battery;
}
int main(int argc,char **argv)
{
ros::init(argc,argv,"third_pkg");
ros::NodeHandle n;    //更新话题的消息格式为自定义的消息格式
```

```
ros::Publisher chatter_pub = n.advertise <third_pkg::myTestMsg>("third_pkg_topic",1000);
ros::ServiceServer service = n.advertiseService("batteryStatus",checkBattery);
ros::Rate loop_rate(2); //发送话题的频率(Hz)0.5 s发送一条,一次电量-1,所以1 s电量-2;
while(ros::ok())
{
battery--;
if(battery <10)battery = 100;
third_pkg::myTestMsg msg; //声明一个自定义消息的对象
msg.name = "corvin";
msg.age = 20;
msg.handsome = true;
msg.salary = 123.45;
chatter_pub.publish(msg); //将消息发布到话题中
ros::spinOnce();
loop_rate.sleep();
}
return 0;
}
```

5. 修改订阅者代码

修改订阅者 subscribe.cpp 代码如下:

```
#include"ros/ros.h"
#include"third_pkg/myTestMsg.h"
#include"third_pkg/myTestSrv.h"
void chatterCallback(const third_pkg::myTestMsg::ConstPtr& msg)
{
ROS_INFO("I heard - name:%s,age:%d,ishandsome:%d,salary:%f",msg->name.c_str(),msg->age,msg->handsome,msg->salary);
}
int main(int argc, char **argv)
{
ros::init(argc,argv,"subscribe_node");
ros::NodeHandle n;
ros::Rate loop_rate(1);
ros::Subscriber sub = n.subscribe("third_pkg_topic",1000,chatterCallback);
ros::ServiceClient client = n.serviceClient <third_pkg::myTestSrv>("batteryStatus");
third_pkg::myTestSrv mySrv;
mySrv.request.index = 1; //发送的服务请求:1,获取电量
while(ros::ok())
{
if(client.call(mySrv))
{
ROS_WARN("Client Get Battery:%d",mySrv.response.result);
}
else
{
ROS_ERROR("Failed to call batteryStatus service");
return 1;
}
ros::spinOnce();
loop_rate.sleep();
}
return 0;
```

}

6. 编译运行查看结果

```
$ cd ~/ros_workspace
$ catkin_make
$ roslaunch third_pkg third_pkg.launch
```

运行结果如图 5.4 所示。

图 5.4 运行结果

课后练习

一、选择题

(1) [单选]下列有关服务与话题通信区别的描述,错误的说法是(　　)。

　　(A) 话题是异步通信,服务是同步通信

　　(B) 多个应答方可以同时提供同一个服务

　　(C) 话题通信是单向的,服务通信是双向的

　　(D) 话题适用于传感器的消息发布,服务适用于偶尔调用的任务

(2) [单选]已知一个服务叫/GetMap,查看该服务的类型可以用的指令是(　　)。

　　(A) rosservice echo /GetMap

　　(B) rosservice type /GetMap

　　(C) rossrv type /GetMap

　　(D) rosservice list /GetMap

(3) [单选]已知/GetMap 的类型是 nav_msgs/GetMap,要查看该类型的具体格式用的指令是(　　)。

　　(A) rossrv show /GetMap

　　(B) rossrv show nav_msgs/GetMap

(C) rosservice show nav_msgs/GetMap

(D) rosservice list nav_msgs/GetMap

(4) [单选]在 parameter server 上添加参数的方式不包括（   ）。

(A) 通过 rosnode 命令添加参数

(B) 通过 rosparam 命令添加参数

(C) 在 launch 中添加参数

(D) 通过 ROS 的 API 来添加参数

二、操作题

编写一个服务控制小乌龟到达一个指定的位置，实现以下功能：

(1) 在 ROS 节点一直保持运行的情况，通过终端不限次数地给定小乌龟目标点；

(2) 小乌龟到达目标点后反馈结果。

第 6 章

参数服务器

ROS 提供了一个参数服务器用于存储一些静态的配置或设置,我们可以方便地从参数服务器中获取、批量下载、修改这些配置。

6.1 roscpp 中的 rosparam

关于参数的 API,roscpp 为我们提供了两套:一套是放在 ros::param namespace 下,另一套是在 ros::NodeHandle 下。这两套 API 的操作完全一样,采用哪一套取决于你的习惯。

6.1.1 getParam()

命令:

getParam("要获取的参数名称",接收参数的变量)

示例:

```
std::string s;
n.getParam("my_param", s);
```

命令解析:

n 为句柄,my_param 为要获取参数的名字,获取到的参数值存到变量 s 中,getParam 返回一个布尔值。该值若为 1,则说明获取参数成功;若为 0,则失效。

```
std::string s;
if (n.getParam("my_param", s))
{
    ROS_INFO("Got param: %s", s.c_str());
}
else
{
    ROS_ERROR("Failed to get param 'my_param'");
}
```

6.1.2 Param()

Param()函数与 get_param 类似都是获取一个参数,不过 param 允许在获取参数时,给定一个默认值,如果获取参数失败则变量中存入所提供的默认值。

命令:

param("参数名称",存储参数的变量,默认值)

示例:

```
int i;
n.param("my_num", i, 42);
```

上述命令将读取参数 my_num 的值存入变量 i,如果读取失败则 i 赋值 42。
针对字符串类型的参数则需要指明数据类型：

```
std::string s;
n.param<std::string>("my_param", s, "default_value");
```

6.1.3 setParam()

setParam()允许用户设置一个参数到参数服务器。

命令：

setParam("参数名称",参数的值)

用法：

```
n.setParam("my_param", "hello there");
```

以上命令将在参数服务中设置一个名字为 my_param 的参数并赋值"hello there"。如果参数服务器上存在 my_param,则直接赋值；如果不存在,则会在参数服务器上创建该参数然后再赋值。

6.1.4 deleteParam()

命令：

deleteParam()

用法：

```
n.deleteParam("my_param");
```

从参数服务器中删除一个参数。

6.1.5 hasParam()

命令：

hasParam()

用法：

```
if (!n.hasParam("my_param"))
{
ROS_INFO("No param named 'my_param'");
}
```

检查参数服务器上是否有该函数。

6.2 通过 launch 加载参数

在 launch 文件中加载参数有以下三种方式：

① 通过 <param> 标签。

<param name = "参数的名字" type = "类型(int bool double str)" value = "参数的值">

② 通过 <rosparam> 标签。

```
<rosparam>
param3: 3
param4: 4
param5: 5
</rosparam>
```

③ 通过加载 yaml 文件。

<rosparam file = " $ (find pkg)/param/costmap_common_params.yaml" command = "load" ns = "global_costmap" />

为了方便说明，接下来创建一个功能包来说明。

```
$ cd ~/ros_workspace/src/
$ catkin_create_pkg rosparamdemo roscpp rospy std_msgs
```

在 rosparamdemo 功能包下新建 launch 文件夹，并在该文件夹下新建一个 param.launch 文件。

在 rosparamdemo 功能包下新建 param 文件夹，并在该文件夹下新建一个 test.yaml 文件。

在 rosparamdemo/src 文件夹下新建一个源文件 param.cpp。

如图 6.1 所示为 rosparamdemo 功能包组成结构。

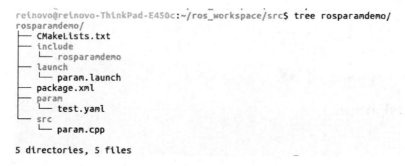

图 6.1　rosparamdemo 功能包组成结构

打开 test.yaml 文件输入以下代码：

```
Pgain3: 3.0
Igain3: 3.0
Dgain3: 3.0
```

打开 param.cpp 文件输入以下代码：

```cpp
#include"ros/ros.h"
int main(int argc,char** argv)
{
  ros::init(argc,argv,"paramdemo");
  ros::NodeHandle n;
```

```cpp
double Pgain1,Igain1,Dgain1;
double Pgain2,Igain2,Dgain2;
double Pgain3,Igain3,Dgain3;
n.getParam("Pgain1",Pgain1);
n.param("Igain1",Igain1,2.0);
ros::param::get("Dgain1",Dgain1);
ROS_INFO("Get Param:P:%f,I:%f,D:%f",Pgain1,Igain1,Dgain1);
n.getParam("Pgain2",Pgain2);
n.param("Igain2",Igain2,2.0);
ros::param::get("Dgain2",Dgain2);
ROS_INFO("Get Param:P:%f,I:%f,D:%f",Pgain2,Igain2,Dgain2);
n.getParam("Pgain3",Pgain3);
n.param("Igain3",Igain3,2.0);
ros::param::get("Dgain3",Dgain3);
ROS_INFO("Get Param:P:%f,I:%f,D:%f",Pgain3,Igain3,Dgain3);
ros::spin();
return 0;
}
```

编辑源码后修改 rosparamdemo 功能包的 CMakeLists.txt，如图 6.2 所示。

```
L35 ## The recommended prefix ensures that target names across packages don't collide
L36  add_executable(${PROJECT_NAME}_node src/param.cpp)
L37
L38 ## Rename C++ executable without prefix
L39 ## The above recommended prefix causes long target names, the following renames the
L40 ## target back to the shorter version for ease of user use
L41 ## e.g. "rosrun someones_pkg node" instead of "rosrun someones_pkg someones_pkg_node"
L42 # set_target_properties(${PROJECT_NAME}_node PROPERTIES OUTPUT_NAME node PREFIX "")
L43
L44 ## Add cmake target dependencies of the executable
L45 ## same as for the library above
L46 # add_dependencies(${PROJECT_NAME}_node ${${PROJECT_NAME}_EXPORTED_TARGETS}
        ${catkin_EXPORTED_TARGETS})
L47
L48 ## Specify libraries to link a library or executable target against
L49  target_link_libraries(${PROJECT_NAME}_node
L50     ${catkin_LIBRARIES}
L51  )
```

图 6.2 rosparamdemo 功能包的 CMakeLists.txt

修改之后编译工作空间。

打开 param.launch 输入以下代码：

```xml
<launch>
<param name = "Pgain1" type = "double" value = "1.0"/>
<param name = "Igain1" type = "double" value = "1.0"/>
<param name = "Dgain1" type = "double" value = "1.0"/>
<rosparam>
  Pgain2: 2.0
  Igain2: 2.0
  Dgain2: 2.0
</rosparam>
<rosparam file = "$(find rosparamdemo)/param/test.yaml" command = "load"/>
<node pkg = "rosparamdemo" type = "rosparamdemo_node" name = "rosparamdemo_node" output = "screen"/>
</launch>
```

运行 param.launch：

```
$ roslaunch rosparamdemo param.launch
```

运行结果如图 6.3 所示。

图 6.3 param.launch 运行结果

6.3 rosparam 命令

ROS 提供了一系列的命令方便我们操作参数服务的参数,如图 6.4 所示。

图 6.4 rosparam 命令

① rosparam list:列出当前参数服务器的参数,如图 6.5 所示。

图 6.5 rosparam list 命令

② rosparam get:获取参数服务器的参数,如图 6.6 所示。

```
reinovo@reinovo-ThinkPad-E450c:~$ rosparam get /Dgain1
1.0
```

图 6.6　rosparam get 命令

③ rosparam set:设置参数服务器的参数,如图 6.7 所示。

```
reinovo@reinovo-ThinkPad-E450c:~$ rosparam set /Dgain1 2.2
reinovo@reinovo-ThinkPad-E450c:~$ rosparam get /Dgain1
2.2
```

图 6.7　rosparam set 命令

注意:当使用 rosparam set 命令时,如果参数服务器上有要设置的参数,则直接修改该参数的数值;如果没有,则创建该参数,并赋值。

④ rosparam load:加载一个 yaml 参数文件。

⑤ rosparam dump:将参数服务器上的参数下载到 yaml 文件中,即

```
$ rosparam dump ~/test.yaml
```

如图 6.8 所示为 test.yaml 文件的内容。

```
reinovo@reinovo-ThinkPad-E450c: ~
Dgain1: 2.2
Dgain2: 2.0
Dgain3: 3.0
Igain1: 1.0
Igain2: 2.0
Igain3: 3.0
Pgain1: 1.0
Pgain2: 2.0
Pgain3: 3.0
rosdistro: 'kinetic

  '
roslaunch:
  uris: {host_reinovo_thinkpad_e450c__46443: 'http://reinovo-ThinkPad-E450c:4644
3/'}
rosversion: '1.12.14

  '
run_id: 987e5f1c-c4ae-11e9-9a3e-507b9d660246
                                                    1,1           All
```

图 6.8　test.yaml 文件内容

6.4　动态参数调节

ROS 中的参数服务器无法在线动态更新,也就是说,如果订阅者不主动查询参数值,就无法获知发布者是否已经修改了参数。这就对 ROS 参数服务器的使用造成了很大的局限,很多场景下我们还是需要动态更新参数的机制,例如参数调试、功能切换等,所以 ROS 提供了另一个非常有用的功能包——dynamic_reconfigure,实现这种动态配置参数的机制。

如图 6.9 所示是启动 Kinect 后 openni 功能包所提供的可动态配置参数的可视化列表。

ROS 中的动态参数修改采用 C/S 架构,在运行过程中,用户在客户端修改参数后不需要重新启动,而是向服务端发送请求,然后服务端通过回调函数确认,即完成参数的动态重配置。

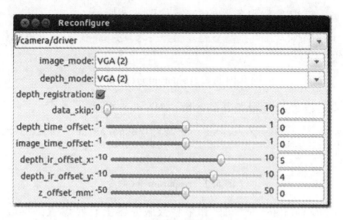

图 6.9 参数动态调节示例

下面就来探索 ROS 中参数动态配置的具体实现方法。

为了方便说明创建一个新的功能包,运行以下代码:

```
$ cd ~/ros_workspace/src/
$ catkin_create_pkg dynamic_tutorials rospy roscpp dynamic_reconfigure
```

实现动态参数配置需要编写一个配置文件,在功能包中创建一个放置配置文件的 cfg 文件夹,然后在其中创建一个配置文件 Tutorials.cfg,输入以下代码:

```
#!/usr/bin/env python
PACKAGE = "dynamic_tutorials"
from dynamic_reconfigure.parameter_generator_catkin import *
gen = ParameterGenerator()
gen.add("int_param",int_t,0,"int parameter",1,0,10);
gen.add("double_param",double_t,0,"double parameter",0.1,0.0,1.0);
gen.add("bool_param",bool_t,0,"bool parameter",True);
gen.add("str_param",str_t,0,"string parameter","ROS_Test1");
size_enum = gen.enum([gen.const("Low",int_t,0,"Low is 0"),
                      gen.const("Medium",int_t,1,"Medium is 1"),
                      gen.const("High",int_t,2,"Hight is 2")],
            "Select from the list")
gen.add("size",int_t,0,"Select from the list",1,0,3,edit_method = size_enum)
exit(gen.generate(PACKAGE,"dynamic_tutorials","Tutorials"))
```

① 导入 dynamic_reconfigure 功能包提供的参数生成器(parameter generator),如图 6.10 所示。

```
#!/usr/bin/env python
PACKAGE = "dynamic_tutorials"

from dynamic_reconfigure.parameter_generator_catkin import *
```

图 6.10 导入参数生成器

② 创建一个参数生成器,如图 6.11 所示。
③ 定义需要动态配置的参数,如图 6.12 所示。

```
gen = ParameterGenerator()
```

图 6.11　创建参数生成器

```
gen.add("int_param", int_t, 0, "An Integer parameter", 50, 0, 100)
gen.add("double_param", double_t, 0, "A double parameter", .5, 0, 1)
gen.add("str_param", str_t, 0, "A string parameter", "Hello World")
gen.add("bool_param",bool_t,0, "A Boolean parameter",  True)
```

图 6.12　定义需要配置的参数

这里定义了四个不同类型的参数,生成参数可以使用参数生成器的 add(name,type,level,description,default,min,max)方法实现,传入参数的意义如下:
- name:参数名,使用字符串描述;
- type:定义参数的类型,可以是 int_t,double_t,str_t,或者 bool_t;
- level:需要传入参数动态配置回调函数中的掩码,在回调函数中会修改所有参数的掩码,表示参数已经修改;
- description:描述参数作用的字符串;
- default:设置参数的默认值;
- min:可选,设置参数的最小值,对于字符串和布尔类型值不生效;
- max:可选,设置参数的最大值,对于字符串和布尔类型值不生效。

④ 这里定义了一个 int_t 类型的参数"size",该参数的值可以通过一个枚举列出来。枚举的定义使用 enum 方法进行定义,其中使用 const 方法定义每一个枚举值的名称、类型、值、描述字符串,如图 6.13 所示。

```
size_enum = gen.enum([gen.const("Small",int_t,0,"A small constant"),
            gen.const("Medium",int_t,1,"A medium constant"),
            gen.const("Large",int_t, 2, "A large constant"),
            gen.const("ExtraLarge",int_t,3,"An extra large constant")],
                "An enum to set size")

gen.add("size", int_t, 0, "A size parameter which is edited via an enum", 1, 0, 3, edit_method=size_enum)
```

图 6.13　定义需要配置的参数

⑤ 最后一行代码用于生成所有 C++和 Python 相关的文件并且退出程序,这里第二个参数表示动态参数运行的节点名,第三个参数是生成文件所使用的前缀,需要和配置文件名相同,如图 6.14 所示。

```
exit(gen.generate(PACKAGE, "dynamic_tutorials", "Tutorials"))
```

图 6.14　关闭退出

配置文件创建完成后,需要使用如下命令为配置文件添加可执行权限:

chmod +x Tutorials.cfg

注意:需要在该配置文件的路径下执行该命令。

⑥ 类似于消息的定义,这里也需要生成代码文件,所以在 CMakeLists.txt 中添加编译规则,如图 6.15 所示。

```
63 ## Generate actions in the 'action' folder
64 # add_action_files(
65 #     FILES
66 #     Action1.action
67 #     Action2.action
68 # )
69 # add dynamic reconfigure api
70   generate_dynamic_reconfigure_options(
71     cfg/Tutorials.cfg
72 )
73 ## Generate added messages and services with any dependencies listed here
74 # generate_messages(
75 #     DEPENDENCIES
76 #     std_msgs  # Or other packages containing msgs
77 # )
```

图 6.15 配置 CMakeLists.txt 文件

⑦ 在 dynamic_tutorials/src 路径下新建 server.cpp 文件输入以下代码:

```
#include "ros/ros.h"
#include <dynamic_reconfigure/server.h>
#include <dynamic_tutorials/TutorialsConfig.h>
void callback(dynamic_tutorials::TutorialsConfig &config, uint32_t level)//回调函数
{
    ROS_INFO("Reconfigure Request: %d %f %s %s %d",
        config.int_param,
        config.double_param,
        config.str_param.c_str(),
        config.bool_param?"True":"False",
        config.size);
}
int main(int argc, char **argv)
{
    ros::init(argc, argv, "node_dynamic_reconfigure");
    dynamic_reconfigure::Server <dynamic_tutorials::TutorialsConfig> server;
                                                                //新建参数服务器
    dynamic_reconfigure::Server <dynamic_tutorials::TutorialsConfig>::CallbackType f;
                                                                //新建参数服务回调函数
    f = boost::bind(&callback, _1, _2);
    server.setCallback(f);
    ros::spin();
    return 0;
}
```

⑧ 修改 CMakeLists.txt 文件添加可执行文件,如图 6.16 所示。

⑨ 编译成功后运行测试效果,如图 6.17 所示。

⑩ 这个时候参数动态配置的服务端就运行起来了,使用 ROS 提供的可视化参数配置工

```
138 ## The recommended prefix ensures that target names across packages don't collide
139 add_executable(dynamic_reconfigure_node |src/server.cpp)
140
141 ## Rename C++ executable without prefix
142 ## The above recommended prefix causes long target names, the following renames the
143 ## target back to the shorter version for ease of user use
144 ## e.g. "rosrun someones_pkg node" instead of "rosrun someones_pkg someones_pkg_node"
145 # set_target_properties(${PROJECT_NAME}_node PROPERTIES OUTPUT_NAME node PREFIX "")
146
147 ## Add cmake target dependencies of the executable
148 ## same as for the library above
149 add_dependencies(dynamic_reconfigure_node ${PROJECT_NAME}_gencfg})
150
151 ## Specify libraries to link a library or executable target against
152 target_link_libraries(dynamic_reconfigure_node
153    ${catkin_LIBRARIES}
154 )
```

图 6.16　CMakeLists.txt 文件修改

```
$ roscore
$ rosrun dynamic_tutorials dynamic_reconfigure_node
```

图 6.17　运行节点

具来修改参数,如图 6.18 所示。

```
$ rosrun rqt_reconfigure rqt_reconfigure
```

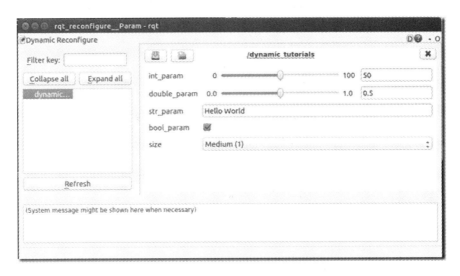

图 6.18　rqt_reconfigure 交互

通过上述界面动态调节参数可以观察终端的数据变化,如图 6.19 所示。

```
[ INFO] [1508144642.464050963]: Reconfigure Request: 50 0.500000 Hello World True 1
[ INFO] [1508144642.466430198]: Spinning node
[ INFO] [1508144747.189317033]: Reconfigure Request: 65 0.500000 Hello World True 1
[ INFO] [1508144752.631543877]: Reconfigure Request: 65 0.230000 Hello World True 1
[ INFO] [1508144757.396002236]: Reconfigure Request: 65 0.230000 hcx True 1
[ INFO] [1508144761.109123375]: Reconfigure Request: 65 0.230000 hcx True 2
[ INFO] [1508144762.807916946]: Reconfigure Request: 65 0.230000 hcx False 2
[ INFO] [1508144763.515548408]: Reconfigure Request: 65 0.230000 hcx True 2
```

图 6.19　终端显示

课后练习

一、选择题

(1) [单选]在 parameter server 上添加参数的方式不包括(　　)。

　　(A) 在 launch 中添加参数

　　(B) 通过 ROS 的 API 来添加参数

　　(C) 通过 rosparam 命令添加参数

　　(D) 通过 rosnode 命令添加参数

(2) [单选]parameter2＝rospy.get_param("/param2"，default＝222)函数语句的功能是(　　)。

　　(A) 在 Parameter Server 上搜索/param2 参数,如果有则将其值存入 parameter2 变量,如果没有则将 222 存入 parameter2 变量

　　(B) 在 Parameter Server 上设置/param2 参数,默认设置为 222

　　(C) 在 Parameter Server 上搜索/param2 参数,将其值存入 parameter2 变量,等待时间为 222 s

　　(D) 在 Parameter Server 上检查/param2 参数,检查其值是否为 222

(3) [多选]ROS 向节点传递参数的方法,以下描述正确的是(　　)。

　　(A) rosrun＋参数服务器传递

　　(B) roslaunch＋参数服务器传递

　　(C) rosrun＋main 参数传递

　　(D) roslaunch＋main 参数传递

二、判断题

rosparam 命令允许在 ROS 的参数服务器上操作和存储数据,参数服务器可以存储整数、浮点数、布尔类型、字典、列表。(　　)

　　(A) 正确

　　(B) 错误

第 7 章
ROS 通信机制——动作

7.1 动作简介

动作库是 ROS 中一个很重要的库,类似服务通信机制,动作库也是一种请求响应机制的通信方式。

动作库主要有以下特点:
① 客户端和服务端通信时,客户端不需要阻塞一直等待服务端完成服务返回。
② 客户端可以随时查询服务端完成服务的进度和状态。
③ 客户端可以随时取消正在执行的服务。

根据以上特点动作库主要应用于以下场景:

当一个服务的执行需要很长时间,或是服务是否能够正确完成存在不确定性,需要随时知道服务的执行进度和状态,可以随时取消服务的场景。如图 7.1 所示的打车经历非常符合动作通信的应用场景。

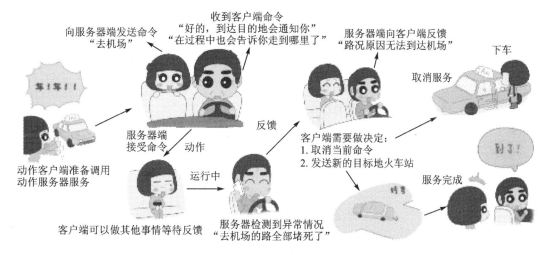

图 7.1 打车服务

动作通信的工作原理是客户端-服务器模式,也是一个双向的通信模式,如图 7.2 所示。通信双方在 ROS 动作协议(Action Protocol)下通过消息进行数据的交流通信。客户端和服务器为用户提供一个简单的 API 来请求目标(在客户端)或通过函数调用和回调来执行目标(在服务器端)。

动作客户端(Action Client)和动作服务器端(Action Server)之间使用动作协议通信。动作协议就是预定义的一组 ROS 消息(message),这些消息被放到 ROS 话题上,在动作客户端和动作服务器端之间进行传输,实现二者的沟通,如图 7.3 所示。

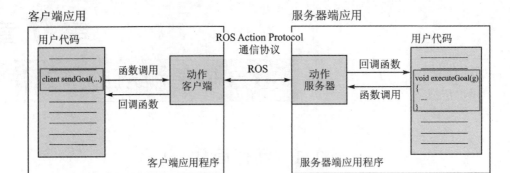

图 7.2　动作通信原理

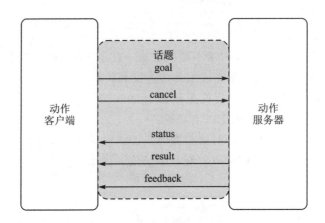

图 7.3　动作客户端和动作服务器端通信内容

图 7.3 中：
- goal　客户端发给服务器的目标；
- cancel　客户端发给服务器的取消任务请求；
- status　发布当前状态下所有目录的状态；
- result　目录完成后服务端向客户端一次性发送的结果内容；
- feedback　服务端定时发布的执行进度信息。

7.2　动作文件规范

类似于服务的创建，动作也需要一个文件用于描述客户端和服务器之间的通信内容的信息格式。该文件通常命名为×××.action 文件，类似于服务的.srv 文件。.action 文件分为三部分：目标(goal)、结果(result)、反馈(feedback)。

（1）目标(goal)

机器人执行一个动作，应该有明确的移动目标信息，包括一些参数的设定，如方向、角度、速度等，从而使机器人完成动作任务。

（2）结果(result)

当运动完成时，动作服务器把本次运动的结果数据发送给客户端，使客户端得到本次动作的全部信息，可能包含机器人的运动时长、最终姿势等。

（3）反馈（feedback）

动作进行的过程中，应该有实时的状态信息反馈给服务器的实施者，告诉实施者动作完成的状态，可以使实施者作出准确的判断去修正命令。

动作规范文件的后缀名是.action，它的内容格式如下：

```
# Define the goal
uint32 dishwasher_id # Specify which dishwasher we want to use
---
# Define the result
uint32 total_dishes_cleaned
---
# Define a feedback message
float32 percent_complete
```

目标、结果、反馈三部分分别用"---"隔开。

7.3 编写一个动作示例

为了大家能够更好地理解动作的原理以及编写方法，下面做一个示例，在该示例中我们创建一个斐波那契数列来演示如何在计算过程中使用动作库。

首先需要认识一下斐波那契数列的特点，这个数列从第三项开始，每一项都等于前两项之和，如图7.4所示。

图7.4 斐波那契数列

动作服务器的设定如下：
- 目标 需要计算的斐波那契数列的长度；
- 反馈 反馈当前正在计算的数列序列；
- 结果 返回指定要求长度而计算得到的数列序列。

① 在ros_worspace中创建功能包，命名为actionlib_test。

$ catkin_create_pkg actionlib_test roscpp rospy std_msgs actionlib message_generation actionlib_msgs

② 在actionlib_test/路径下新建action文件夹，并在该文件夹下创建Fibonacci.action文件，如图7.5所示。

③ 修改CMakeLists.txt以及package.xml文件如图7.6和图7.7所示。

④ 编译、输入以下命令编译工作空间：

$ cd ~/ros_workspace
$ catkin_make

正常情况下将在ros_workspace/devel/include/actionlib_test路径下生成如图7.8所示的内容。

```
reinovo@reinovo-ThinkPad-E450c:~/ros_workspace/src$ tree actionlib_test/
actionlib_test/
├── action
│   └── Fibonacci.action
├── CMakeLists.txt
├── include
│   └── actionlib_test
├── package.xml
└── src
    ├── client.cpp
    └── server.cpp
```

```
int32 order
---
int32[] result_sequence
---
int32[] feedback_sequence
~
~
```

图 7.5　创建文件

```
## Generate actions in the 'action' folder
add_action_files(
  DIRECTORY action
  FILES Fibonacci.action
)

## Generate added messages and services with any dependencies listed here
generate_messages(
  DEPENDENCIES
  actionlib_msgs std_msgs
)

catkin_package(
#  INCLUDE_DIRS include
#  LIBRARIES actionlib_test
   CATKIN_DEPENDS actionlib actionlib_msgs message_generation roscpp rospy std_msgs
#  DEPENDS system_lib
)
```

图 7.6　CMakeLists.txt 文件修改

```xml
<buildtool_depend>catkin</buildtool_depend>
<build_depend>actionlib</build_depend>
<build_depend>actionlib_msgs</build_depend>
<build_depend>message_generation</build_depend>
<build_depend>roscpp</build_depend>
<build_depend>rospy</build_depend>
<build_depend>std_msgs</build_depend>
<build_export_depend>actionlib</build_export_depend>
<build_export_depend>actionlib_msgs</build_export_depend>
<build_export_depend>roscpp</build_export_depend>
<build_export_depend>rospy</build_export_depend>
<build_export_depend>std_msgs</build_export_depend>
<exec_depend>actionlib</exec_depend>
<exec_depend>actionlib_msgs</exec_depend>
<exec_depend>roscpp</exec_depend>
<exec_depend>rospy</exec_depend>
<exec_depend>std_msgs</exec_depend>
<exec_depend>message_generation</exec_depend>
```

图 7.7　package.xml 文件修改

⑤ 编写源码，在 actionlib_test/src 路径下新建两个源文件 server.cpp 以及 client.cpp。server.cpp 源码：

图 7.8 系统生成 action 头文件

```
#include <ros/ros.h>
#include <actionlib/server/simple_action_server.h>
//包含动作库提供的头文件,用来实现简单的动作
#include <actionlib_test/FibonacciAction.h>
//包含自定义的头文件,由 FibonacciAction.msg 自动生成
class FibonacciAction
{
protected:
ros::NodeHandle nh_;//nodehandle 的实例化必须在动作服务器的前面
actionlib::SimpleActionServer <actionlib_test::FibonacciAction> as_;
//利用动作的 roscpp 接口新建一个动作服务器
std::string action_name_;
//创建消息被用来发布反馈结果
actionlib_test::FibonacciFeedback feedback_;
actionlib_test::FibonacciResult result_;
public:
//构造函数,创建动作服务器并绑定回调函数
FibonacciAction(std::string name):
as_(nh_,name,boost::bind(&FibonacciAction::executeCB,this,_1),false),
action_name_(name)//定义构造函数时分别对 as_,action_name_初始化
{as_.start();}
~FibonacciAction(void)
{}
void executeCB(const actionlib_test::FibonacciGoalConstPtr &goal)
{
ros::Rate r(1);
bool success = true;
//在数列中放两个值 0 和 1
feedback_.feedback_sequence.clear();
feedback_.feedback_sequence.push_back(0);
feedback_.feedback_sequence.push_back(1);
ROS_INFO("%s:executing,order:%i with seeds %i,%i",action_name_.c_str(),
goal->order,feedback_.feedback_sequence[0],feedback_.feedback_sequence[1]);
//开始执行动作,这里通过一个 for 循环来模拟长时间的执行任务
for(int i = 1;i <= goal->order;i++)
{
//在执行任务的过程中,还可以接收客户端的 cancel 命令,这样就设置状态 preempted
if(as_.isPreemptRequested()||! ros::ok())
{
ROS_INFO("%s:Preempted",action_name_.c_str());
as_.setPreempted();
success = false;
```

```cpp
            break;
        }
        feedback_.feedback_sequence.push_back(feedback_.feedback_sequence[i] + feedback_.feedback_sequence[i-1]);
        as_.publishFeedback(feedback_);
        r.sleep();
    }
    //判断最终的任务执行状态是否成功,成功则设置succeed
    if(success)
    {
        result_.result_sequence = feedback_.feedback_sequence;
        ROS_INFO("%s:Succeeded",action_name_.c_str());
        as_.setSucceeded(result_);
    }
  }
};
//main函数,创建动作并启动,节点spin循环,然后等待接收客户端发送的目标来执行任务
int main(int argc,char** argv)
{
    ros::init(argc,argv,"server");
    FibonacciAction fibonacci("fibonacci");
    ros::spin();
    return 0;
}
```

client.cpp 源码:

```cpp
#include <ros/ros.h>
#include <actionlib/client/simple_action_client.h>
#include <actionlib/client/terminal_state.h>  //包含目标可能用的一些状态
#include <actionlib_test/FibonacciAction.h>
int main(int argc, char** argv)
{
    ros::init(argc,argv,"client");
    actionlib::SimpleActionClient<actionlib_test::FibonacciAction> ac("fibonacci",true);
    ROS_INFO("Waiting for action server to start.");
    ac.waitForServer();
    ROS_INFO("Action server started,sending goal.");
    actionlib_test::FibonacciGoal goal;
    goal.order = 20;
    ac.sendGoal(goal);
    bool timeout = ac.waitForResult(ros::Duration(30.0));
    if(timeout)
    {
        actionlib::SimpleClientGoalState state = ac.getState();
        ROS_INFO("Action finished:%s",state.toString().c_str());
    }
    else{
        ROS_INFO("Action did not finish before the timeout.");
    }
    return 0;
}
```

⑥ 修改 CMakeLists.txt 文件并编译工作空间,如图 7.9 所示。

```
 add_executable(server_node src/server.cpp)
 add_executable(client_node src/client.cpp)

## Rename C++ executable without prefix
## The above recommended prefix causes long target names, the following renames the
## target back to the shorter version for ease of user use
## e.g. "rosrun someones_pkg node" instead of "rosrun someones_pkg someones_pkg_node"
# set_target_properties(${PROJECT_NAME}_node PROPERTIES OUTPUT_NAME node PREFIX "")

## Add cmake target dependencies of the executable
## same as for the library above
add_dependencies(server_node ${${PROJECT_NAME}_EXPORTED_TARGETS} ${catkin_EXPORTED_TARGETS})
add_dependencies(client_node ${${PROJECT_NAME}_EXPORTED_TARGETS} ${catkin_EXPORTED_TARGETS})


## Specify libraries to link a library or executable target against
 target_link_libraries(server_node
   ${catkin_LIBRARIES}
 )
 target_link_libraries(client_node
   ${catkin_LIBRARIES}
 )
```

图 7.9 CMakeLists.txt 文件修改

注意：

```
add_dependencies(server_node ${${PROJECT_NAME}_EXPORTED_TARGETS} ${catkin_EXPORTED_TARGETS})
add_dependencies(client_node ${${PROJECT_NAME}_EXPORTED_TARGETS} ${catkin_EXPORTED_TARGETS})
```

当源码里面包含了一些自定义头文件或是自定义消息服务时，在这些头文件生成前需要编写源码时，要打开上面部分。

修改之后，编译工作空间。

⑦ 运行节点测试：

运行 ROS 节点管理器：

```
$ roscore
```

运行客户端：

```
$ rosrun actionlib_test client_node
```

如图 7.10 所示，客户端启动后会一直等待服务器端开启。

```
reinovo@reinovo-ThinkPad-E450c: ~
reinovo@reinovo-ThinkPad-E450c:~$ rosrun actionlib_test client_node
[ INFO] [1567599900.828555780]: Waiting for action server to start.
```

图 7.10 客户端运行

运行服务器端：

```
$ rosrun actionlib_test server_node
```

如图 7.11 所示，服务器开启后会收到客户端的请求。

```
reinovo@reinovo-ThinkPad-E450c: ~
reinovo@reinovo-ThinkPad-E450c:~$ rosrun actionlib_test server_node
[ INFO] [1567600095.678374515]: fibonacci:executing,order:20 with seeds 0,1
```

图 7.11 服务器端运行

如图 7.12 所示当服务器端完成服务后,向客户端发送成功信号。

图 7.12　服务成功完成

在服务执行过程中的话题列表,如图 7.13 所示。

图 7.13　话题列表

订阅话题/fibonacci/feedback 可以随时查看任务执行的进度,如图 7.14 所示。

图 7.14　/fibonacci/feedback 话题内容

课后练习

一、选择题

(1)[单选]关于动作的描述错误的是(　　)。

　　(A) 动作通信的双方也是客户端和服务器

　　(B) 动作的客户端可以发送目标,也可以请求取消

　　(C) .action 文件与.srv 文件写法一致

　　(D) 动作通常用在长时间的任务中

（2）［多选］关于 ROS 通信方式的描述正确的是（　　）。
　　（A）现在要设计一个节点,开发路径规划功能,输入是目标点和起始点,输出是路径,
　　　　 适合用话题通信方式
　　（B）传感器消息发布一般都采用话题形式发布
　　（C）动作更适合用在执行时间长并且需要知道状态和结果的场景
　　（D）机械臂关节逆解适合用服务通信

二、操作题

编写一个动作控制一个小乌龟顺序到达一系列点,实现以下功能：
（1）在 ROS 节点一直保持运行的情况,通过终端不限次数地给定小乌龟目标点；
（2）小乌龟到达目标点后反馈结果。

第 8 章
什么是 tf

8.1 tf 介绍

机器人的坐标变换一直以来是机器人学习的一个难点,人类在进行一个简单的动作时,从思考到实施行动再到完成动作可能仅仅需要几秒钟,但是对于机器人来说就需要进行大量的计算和坐标转换。

观察图 8.1 所示的这个机器人,直观上我们不认为拿起物品会有什么难度,站在人类的立场上,也许会想到手向前伸—抓住—手收回,就完成了这一系列的动作。但是机器人远远没有这么智能,它能得到的只是各种传感器发送回来的数据,然后它再处理各种数据进行操作,比如手臂弯曲 45°,再向前移动 20 cm 等各种十分精确的数据,尽管如此,机器人依然无法做到像人类一样自如地进行各种行为操作。那么在这个过程中,tf 又扮演着什么样的角色呢?还拿图 8.1 来说,机器人的"眼睛"获取一组关于物体坐标方位的数据,而相对于机器人手臂来说,这个坐标只是相对于机器人头部的传感器,并

图 8.1 机器人抓取目标物体

不直接适用于机器人手臂执行,那么物体相对于头部和手臂之间的坐标转换,就是 tf。坐标变换包括了位置和姿态两个方面的变换,ROS 中的 tf 是一个可以让用户随时记录多个坐标系的软件包。tf 保持缓存的树形结构中的坐标系之间的关系,并且允许用户在任何期望的时间点、在任何两个坐标系之间转换点、矢量等。

总之,tf 是一个用户随时间跟踪多个坐标系的包,tf 管理一系列树状结构坐标系之间的关系。允许用户在各个坐标系中进行点、向量的变换。通俗地说,tf 可以帮助我们实时地在各个坐标系中进行坐标转换。

8.2 tf 示例

8.2.1 示例运行

运行以下命令安装该教程所需的插件:

```
$ sudo apt-get install ros-kinetic-ros-tutorials ros-kinetic-geometry-tutorials ros-
kinetic-rviz ros-kinetic-rosbash ros-kinetic-rqt-tf-tree
```

运行以下命令执行 demo 示例，运行结果如图 8.2 所示。

```
$ roslaunch turtle_tf turtle_tf_demo.launch
```

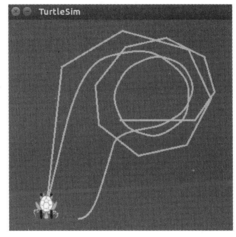

图 8.2 turtle_tf_demo 运行结果

将当前运行命令的窗口作为激活窗口，用箭头按键控制小乌龟移动，另一只小乌龟将会跟着移动。

这个 demo 用 tf 库建立了三个坐标系：世界坐标系、乌龟 1 坐标系、乌龟 2 坐标系。在乌龟移动的过程中，tf 库创建了一个 tf 发布器，用于发布乌龟 1 的坐标系在世界坐标系中的实时位置，然后利用一个 tf 接收器，接收该位置，然后计算两个乌龟坐标系的差异，让乌龟 2 跟着乌龟 1 移动。

8.2.2 tf 命令工具

1. view_frames 命令

view_frames 命令会以一个树状图的形式，给我们展现出当前 ROS 中 tf 发布的所有坐标系。打开一个终端输入以下命令：

```
$ rosrun tf view_frames
```

图 8.3 中所示生成了一个命名为 frames.pdf 的文件。

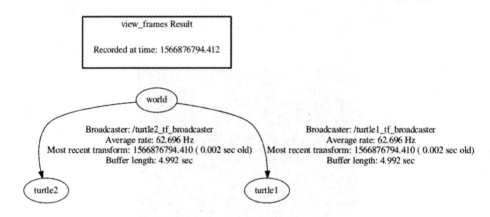

图 8.3 命令运行结果

运行以下命令查看 pdf 文件内容，如图 8.4 所示。

```
$ evince frames.pdf
```

图 8.4 frames.pdf 文件内容

图 8.4 为 tf 树，每一个椭圆都是一个帧，联通的两个帧之间存在一个广播节点，该节点负责实时发布两个帧之间的转换关系。

2. rqt_tf_tree 命令

rqt_tf_tree 命令是一个可以实时可视化显示当前 tf 发布的坐标系图，效果同图 8.4 中的效果。

```
$ rosrun rqt_tf_tree rqt_tf_tree
```

图 8.5 所示为运行结果。

3. tf_echo 命令

tf_echo 命令可以报告任意两个坐标系之间的变换。

用法：

rosrun tf tf_echo [reference_frame] [targe_frame]

打开终端输入以下命令：

```
$ rosrun tf tf_echo turtle1 turtle2
```

什么是 tf

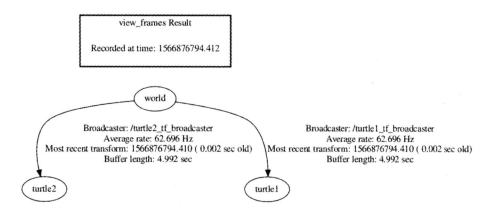

图 8.5 rqt_tf_tree 命令结果

以 turtle1 为参考坐标系 turtle2 的转换关系如图 8.6 所示。

```
retnovo@retnovo-ThinkPad-E450c:~$ rosrun tf tf_echo turtle2 turtle1
At time 1566877689.578
- Translation: [0.000, 0.000, 0.000]
- Rotation: in Quaternion [0.000, 0.000, -0.352, 0.936]
            in RPY (radian) [0.000, 0.000, -0.719]
            in RPY (degree) [0.000, 0.000, -41.182]
At time 1566877690.346
- Translation: [0.000, 0.000, 0.000]
- Rotation: in Quaternion [0.000, 0.000, -0.352, 0.936]
            in RPY (radian) [0.000, 0.000, -0.719]
            in RPY (degree) [0.000, 0.000, -41.182]
At time 1566877691.338
- Translation: [0.000, 0.000, 0.000]
- Rotation: in Quaternion [0.000, 0.000, -0.352, 0.936]
            in RPY (radian) [0.000, 0.000, -0.719]
            in RPY (degree) [0.000, 0.000, -41.182]
At time 1566877692.346
- Translation: [0.000, 0.000, 0.000]
- Rotation: in Quaternion [0.000, 0.000, -0.352, 0.936]
            in RPY (radian) [0.000, 0.000, -0.719]
            in RPY (degree) [0.000, 0.000, -41.182]
```

图 8.6 以 turtle1 为参考坐标系 turtle2 的转换关系

8.2.3 tf 中的消息

在 8.2.2 小节中介绍的 ROS 中的 tf 已经使我们初步认识了 tf 和 tf 树,了解了在每个帧之间都会由广播来发布消息维系坐标转换,那么这个消息到底是什么样子的呢？这个消息 TransformStampde.msg,就是处理两个帧之间一小段 tf 的数据格式。

TransformStamped.msg 的格式规范如下：

```
std_mags/Header header
uint32 seq
time stamp
string frame_id
string child_frame_id
geometry_msgs/Transform transform
geometry_msgs/Vector3 translation
float64 x
float64 y
float64 z
```

```
geometry_msgs/Quaternion rotation
    float64 x
    float64 y
    flaot64 z
    float64 w
```

观察标准的格式规范，首先 header 定义了序号、时间以及帧的名称；接着还写了 child_frame，这两个帧之间要做那种变换就是由 geometry_msgs/Transform 来定义的。

Vector3 三维向量表示平移，Quaternion 四元数表示旋转。如图 8.5 所示的 tf 树中的 world 和 turtle1 两个帧之间的消息，就是由这种格式来定义的。world 就是 frame_id,turtle1 就是 child_frame_id。我们知道，一个话题可能会有很多个节点向上面发送消息。最终，许多的 TransformStamped.msg 发向 tf,形成了 tf 树。

tf 树的消息格式：tf2_msgs/TFMessage。

```
geometry_msgs/TransformStamped[] transforms
    std_msgs/Header header
        uint32 seq
        time stamp
        string frame_id
    string child_frame_id
    geometry_msgs/Transform transform
        geometry_msgs/Vector3 translation
            float64 x
            float64 y
            float64 z
        geometry_msgs/Quaternion rotation
            float64 x
            float64 y
            flaot64 z
            float64 w
```

如上所述，一个 TransformStamped 数组就是一个 tf 树。

8.3　tf 的 C++ 接口

前面介绍了 tf 的基本概念和 tf 树消息的格式类型。我们知道，tf 不仅仅是一个标准、话题，它还是一个接口。本节我们就介绍 C++ 中 tf 的一些函数和写法。

8.3.1　数据类型

C++ 中给我们提供了很多 tf 的数据类型，如表 8.1 所列。

注意：虽然此表的最后"带时间戳的变换"数据类型为 tf::StampedTransform,与 8.2 节所讲的 geometry_msgs/TransformStamped.msg 看起来很相似，但其实数据类型完全不一样，tf::StampedTransform 只能用在 C++ 里，只是 C++ 的一个类、一种数据格式，并不是一个消息。而 geometry_msgs/TransformStamped.msg 是一个消息，它依赖于 ROS,与语言无关，亦即无论何种语言，C++、Python、Java 等，都可以发送该消息。

表 8.1　tf 数据类型

名　称	数据类型
向量	tf::Vector3
点	tf::Point
四元数	tf:Quaternion
3×3矩阵（旋转矩阵）	tf::Matrix3x3
位姿	tf:pose
变换	tf::Transform
带时间戳的以上类型	tf::Stamped
带时间戳的变换	tf::StampedTransform

8.3.2　数据转换

在 tf 里有可能会遇到各种数据的转换,例如常见的四元数、旋转矩阵、欧拉角这三种数据之间的转换,如图 8.7 所示。tf in roscpp 给了我们解决该问题的函数。首先在 tf 中与数据转化的数据的类型都包含在 #include <tf/tf.h> 头文件中。

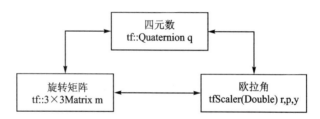

图 8.7　常见的数据转换

定义空间点和空间向量如表 8.2 所列。

表 8.2　定义空间点和空间向量

编　号	函数名称	函数功能
1.1	tfScalar::tfDot(const Vector3 &v1, const Vector3 &v2)	计算两个向量的点积
1.2	tfScalar length()	计算向量的模
1.3	Vector3 &normalize()	求与已知向量同方向的单位向量
1.4	tfScalar::tfAngle(const Vector3 &v1, const Vector3 &v2)	计算两个向量的夹角
1.5	tfScale:tfDistance(const Vector3 &v1, constVector3 &v2)	计算两个向量的距离
1.6	tfScale::tfCross(const Vector3 &v1, const Vector3&v2)	计算两个向量的乘积

示例 1 代码如下：

```
#include <ros/ros.h>
#include <tf/tf.h>
int main(int argc, char** argv){
//初始化
```

```cpp
ros::init(argc, argv, "coordinate_transformation");
ros::NodeHandle node;
tf::Vector3 v1(1,1,1);
tf::Vector3 v2(1,0,1);//第1部分,定义空间点和空间向量
std::cout << "第1部分,定义空间点和空间向量" << std::endl;
//1.1 计算两个向量的点积
std::cout << "向量v1:" << "(" << v1[0] << "," << v1[1] << "," << v1[2] << "),";
std::cout << "向量v2:" << "(" << v2[0] << "," << v2[1] << "," << v2[2] << ")" << std::endl;
std::cout << "两个向量的点积:" << tfDot(v1,v2) << std::endl;
//1.2 计算向量的模
std::cout << "向量v2的模值:" << v2.length() << std::endl;
//1.3 求与已知向量同方向的单位向量
tf::Vector3 v3;
v3 = v2.normalize();
std::cout << "与向量v2的同方向的单位向量v3:" << "(" << v3[0] << "," << v3[1] << "," << v3[2] << ")" << std::endl;
//1.4 计算两个向量的夹角
std::cout << "两个向量的夹角(弧度):" << tfAngle(v1,v2) << std::endl;
//1.5 计算两个向量的距离
std::cout << "两个向量的距离:" << tfDistance2(v1,v2) << std::endl;
//1.6 计算两个向量的乘积
tf::Vector3 v4;
v4 = tfCross(v1,v2);
std::cout << "两个向量的乘积v4:" << "(" << v4[0] << "," << v4[1] << "," << v4[2] << ")" << std::endl;
return 0;
}
```

定义四元数如表8.3所列。

表8.3 定义四元数

编 号	函数名称	函数功能
2.1	setRPY(const tfScalar& yaw, const stScalar &pitch,const tfScalar &roll)	由欧拉角计算四元数
2.2	Vector3 getAxis()	由四元数得到旋转轴
2.3	setRotation(const Vector3 &axis, const tfScalar& angle)	由旋转轴和旋转角估计四元数

示例2代码如下:

```cpp
#include <ros/ros.h>
#include <tf/tf.h>
int main(int argc, char** argv){
//初始化
ros::init(argc, argv, "coordinate_transformation");
ros::NodeHandle node;
std::cout << "第2部分,定义四元数" << std::endl;
//2.1 由欧拉角计算四元数
tfScalar yaw,pitch,roll;
yaw = 0;pitch = 0;roll = 0;
std::cout << "欧拉角rpy(" << roll << "," << pitch << "," << yaw << ")";
tf::Quaternion q;
q.setRPY(yaw,pitch,roll);
```

```
        std::cout << ",转化到四元数 q:" << "(" << q[3] << "," << q[0] << "," << q[1] << "," << q[2] << ")"
<< std::endl;
    //2.2 由四元数得到旋转轴
    tf::Vector3 v5;
    v5 = q.getAxis();
    std::cout << "四元数 q 的旋转轴 v5" << "(" << v5[0] << "," << v5[1] << "," << v5[2] << ")" << std::
endl;
    //2.3 由旋转轴和旋转角估计四元数
    tf::Quaternion q2;
    q2.setRotation(v5,1.570796);
    std::cout << "旋转轴 v5 和旋转角 90 度,转化到四元数 q2:" << "(" << q2[3] << "," << q2[0] << ",
" << q2[1] << "," << q2[2] << ")" << std::endl;
    return 0;
}
```

定义旋转矩阵如表 8.4 所列。

表 8.4 定义旋转矩阵

编 号	函数名称	函数功能
3.1	setRotaion(const Quaternion &q)	通过四元数得到旋转矩阵
3.2	getEulerYPR(tfScalar &yaw, tfScalar &pitch, tfScalar &roll)	由旋转矩阵求欧拉角

示例 3 代码如下：

```
#include <ros/ros.h>
#include <tf/tf.h>
int main(int argc, char** argv){
    //初始化
    ros::init(argc, argv, "coordinate_transformation");
    ros::NodeHandle node;
    //第 3 部分,定义旋转矩阵
    std::cout << "第 3 部分,定义旋转矩阵" << std::endl;
    //3.1 通过四元数得到旋转矩阵
    tf::Quaternion q2(1,0,0,0);
    tf::Matrix3x3 Matrix;
    tf::Vector3 v6,v7,v8;
    Matrix.setRotation(q2);
    v6 = Matrix[0];
    v7 = Matrix[1];
    v8 = Matrix[2];
    std::cout << "四元数 q2 对应的旋转矩阵 M:" << v6[0] << "," << v6[1] << "," << v6[2] << std::endl;
    std::cout << " " << v7[0] << "," << v7[1] << "," << v7[2] << std::endl;
    std::cout << " " << v8[0] << "," << v8[1] << "," << v8[2] << std::endl;
    //3.2 由旋转矩阵求欧拉角
    tfScalar m_yaw,m_pitch,m_roll;
    Matrix.getEulerYPR(m_yaw,m_pitch,m_roll);
    std::cout << "由旋转矩阵 M,得到欧拉角 rpy(" << m_roll << "," << m_pitch << "," << m_yaw << ")"
<< std::endl;
    return 0;
};
```

示例 4:欧拉角转化成四元数。代码如下：

```cpp
#include <ros/ros.h>
#include <tf/tf.h>
int main(int argc, char** argv){
//初始化
ros::init(argc, argv, "Euler2Quaternion");
ros::NodeHandle node;
geometry_msgs::Quaternion q;
double roll,pitch,yaw;
while(ros::ok())
{
//输入一个相对原点的位置
std::cout << "输入的欧拉角:roll,pitch,yaw:";
std::cin >> roll >> pitch >> yaw;
//输入欧拉角,转化成四元数在终端输出
q = tf::createQuaternionMsgFromRollPitchYaw(roll,pitch,yaw);
//ROS_INFO("输出的四元数为:w=%d,x=%d,y=%d,z=%d",q.w,q.x,q.y,q.z);
std::cout << "输出的四元数为:w=" << q.w << ",x=" << q.x << ",y=" << q.y << ",z=" << q.z << std::endl;
ros::spinOnce();
}
return 0;
};
```

示例5:四元数转化成欧拉角。代码如下:

```cpp
#include <ros/ros.h>
#include "nav_msgs/Odometry.h"
#include <tf/tf.h>
int main(int argc, char** argv){
//初始化
ros::init(argc, argv, "Quaternion2Euler");
ros::NodeHandle node;
nav_msgs::Odometry position;
tf::Quaternion RQ2;
double roll,pitch,yaw;
while(ros::ok())
{
//输入一个相对原点的位置
std::cout << "输入的四元数:w,x,y,z:";
std::cin >> position.pose.pose.orientation.w >> position.pose.pose.orientation.x >> position.pose.pose.orientation.y >> position.pose.pose.orientation.z;
//输入四元数,转化成欧拉角在终端输出
tf::quaternionMsgToTF(position.pose.pose.orientation,RQ2);
// tf::Vector3 m_vector3;方法2
// m_vector3 = RQ2.getAxis();
tf::Matrix3x3(RQ2).getRPY(roll,pitch,yaw);
std::cout << "输出的欧拉角为:roll=" << roll << ",pitch=" << pitch << ",yaw=" << yaw << std::endl;
//std::cout << "输出的欧拉角为:roll=" << m_vector3[0] << ",pitch=" << m_vector3[1] << ",yaw=" << m_vector3[2] << std::endl;
ros::spinOnce();
}
return 0;
};
```

8.3.3　tf 类

● tf::TransformBroadcaster 类

```
transformBroadcaster()
void sendTransform(const StampedTransform &transform)
void sendTransform(const std::vector <StampedTransform> &transforms)
void sendTransform(const geometry_msgs::TransformStamped &transform)
void sendTransform(const std::vector <geometry_msgs::TransformStamped> &transforms)
```

这个类在前面讲 tf 树时提到过,这个广播就是一个发布者,而 sendTransform 的作用是来封装发布的函数。在实际使用中,我们需要在某个节点中构建 tf::TransformBroadcaster 类,然后调用 sendTransform(),将 transform 发布到 /tf 的一段 transform 上。/tf 里的 transform 为我们重载了多种不同的函数类型。

示例 6 代码如下:

```cpp
#include <ros/ros.h>
#include <tf/transform_broadcaster.h>
#include <tf/tf.h>
int main(int argc, char** argv){
//初始化
ros::init(argc, argv, "tf_broadcaster");
ros::NodeHandle node;
static tf::TransformBroadcaster br;
tf::Transform transform;
//geometry_msgs::Quaternion qw;
tf::Quaternion q;
//定义初始坐标和角度
double roll = 0,pitch = 0,yaw = 0,x = 1.0,y = 2.0,z = 3.0;
ros::Rate rate(1);
while(ros::ok())
{
yaw += 0.1;//每经过 1 s 开始一次变换
//输入欧拉角,转化成四元数在终端输出
q.setRPY(roll,pitch,yaw);
//qw = tf::createQuaternionMsgFromRollPitchYaw(roll,pitch,yaw);方法 2
transform.setOrigin(tf::Vector3(x,y,z));
transform.setRotation(q);
std::cout << "发布 tf 变换:sendTransform 函数" << std::endl;
br.sendTransform(tf::StampedTransform(transform,ros::Time::now(),"base_link","link1"));
std::cout << "输出的四元数为:w = " << q[3] << ",x = " << q[0] << ",y = " << q[1] << ",z = " << q[2] << std::endl;
// std::cout << "输出的四元数为:w = " << qw.w << ",x = " << qw.x << ",y = " << qw.y << ",z = " << qw.z << std::endl;
rate.sleep();
ros::spinOnce();
}
return 0;
};
```

● tf::TransformListener 类

```
void lookupTranform(const std::string &target_frame, const std::string &source_frame, const
```

```
ros::Time &time, StampedTransform &transform)const
    bool canTransform()
    bool waitForTransform()const
```

上一个类是向/tf 上发的类,那么这一个就是从/tf 上接收的类。首先看 lookupTransform()函数,第一个参数是目标坐标系;第二个参数为源坐标系,也即得到从源坐标系到目标坐标系之间的转换关系;第三个参数为查询时刻;第四个参数为存储转换关系的位置。值得注意是,第三个参数通常用 ros::Time(0),这个表示为最新的坐标转换关系,而 ros::time::now 则会因为收发延迟的原因,而不能正确获取当前最新的坐标转换关系。canTransform()是用来判断两个转换之间是否连通,waitForTransform()const 是用来等待某两个转换之间的连通。

示例 7 代码如下:

```cpp
#include <ros/ros.h>
#include <tf/transform_listener.h>
#include <geometry_msgs/Twist.h>
int main(int argc, char** argv){
ros::init(argc, argv, "tf_listener");
ros::NodeHandle node;
tf::TransformListener listener;
//1. 阻塞直到帧相通
std::cout << "1. 阻塞直到帧相通" << std::endl;
listener.waitForTransform("/base_link","link1",ros::Time(0),ros::Duration(4.0));
ros::Rate rate(1);
while (node.ok()){
tf::StampedTransform transform;
try{
//2. 监听对应的 tf,返回平移和旋转
std::cout << "2. 监听对应的 tf,返回平移和旋转" << std::endl;
listener.lookupTransform("/base_link", "/link1",
ros::Time(0), transform);
//ros::Time(0)表示最近的一帧坐标变换,不能写成 ros::Time::now()
}
catch (tf::TransformException &ex) {
ROS_ERROR("%s",ex.what());
ros::Duration(1.0).sleep();
continue;
}
std::cout << "输出的位置坐标:x = " << transform.getOrigin().x() << ",y = " << transform.getOrigin().y() << ",z = " << transform.getOrigin().z() << std::endl;
std::cout << "输出的旋转四元数:w = " << transform.getRotation().getW() << ",x = " << transform.getRotation().getX() << ",y = " << transform.getRotation().getY() << ",z = " << transform.getRotation().getZ() << std::endl;
rate.sleep();
}
return 0;
}
```

课后练习

一、选择题

(1) [单选]tf 在 ROS 中的抽象数据结构是（　　）。
  (A) 树　　　(B) 链表　　　(C) 队列　　　(D) 栈

(2) [单选]查看当前从"map"坐标系和"right_wheel"坐标系之间的变换关系,应该用 C++的哪句指令？（　　）
  (A) lookupTransform("map", "right_wheel", ros::Time(0), trans_result);
  (B) getTransform("map", "right_wheel", ros::Time::now(), trans_result);
  (C) getTransform("map", "right_wheel", ros::Time(0), trans_result);
  (D) lookupTransform("map", "right_wheel", ros::Time::now(), trans_result);

(3) [单选]图形化查看当前的 tf 树可以用的指令是（　　）。
  (A) $ rosrun rqt_tf_tree rqt_tf_tree
  (B) $ rosrun tf tf_echo
  (C) $ rviz
  (D) $ rosrun tf display

(4) [单选]/tf 话题在 ROS 中可能的消息类型是（　　）。
  (A) tf/tfMessage
  (B) tf2_msgs/TFMessage
  (C) tf/tfmsg
  (D) tf2/TFMessage

二、判断题

URDF 文件包括了 link 和 joint 两个主要标签,前者定义机器人的零部件（如形状、坐标）,后者定义零部件之间的运动关系（如旋转、平移）。
(A) 正确
(B) 错误

三、操作题

创建一个功能包,实现三只小乌龟能够以等边三角形队形移动,完成以下指标:
(1) 可以用键盘控制领头的小乌龟运动,其他小乌龟跟随领头小乌龟运动,期间保持等边三角形队形;
(2) 等边三角形的边长可以在 launch 文件中指定;
(3) 所有节点在一个 launch 文件中启动;
(4) 默认等边三角形的边长为 1。

第 9 章
ROS 车型机器人建模

在本章中首先介绍机器人的系统架构,然后借助 ROS 提供的工具创建一个车型虚拟机器人。

9.1 机器人组成架构

从控制的角度分析,一个机器人由四部分组成:执行机构、驱动系统、传感系统、控制系统,如图 9.1 所示。

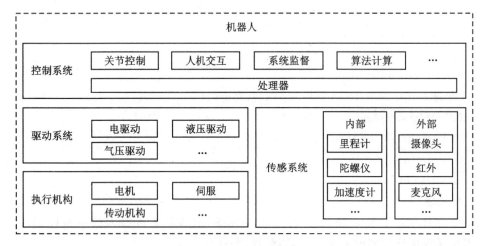

图 9.1 机器人组成架构

① 执行机构:人体的手和脚,直接面向工作对象的机械装置。

② 驱动系统:人体的肌肉和筋络,负责驱动执行机构,将控制系统下达的命令转换成执行机构需要的信号。

③ 传感系统:人体的感官和神经,主要完成信号的输入和反馈,包括内部传感系统和外部传感系统。

④ 控制系统:人体的大脑,实现任务及信息的处理,输出控制命令信号。

机器人控制回路如图 9.2 所示。

如图 9.3 所示为 REI_bobac 教育机器人的结构组成。

下面以 bobac 机器人为例讲解机器人的架构。

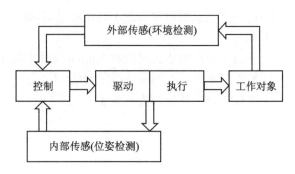

图 9.2 机器人控制回路

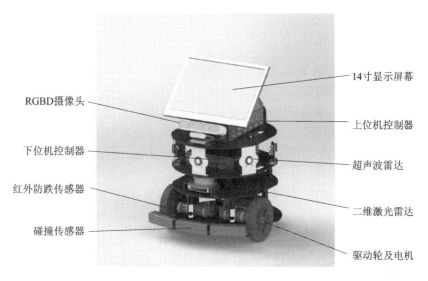

图 9.3　REI_bobac 教育机器人的结构

9.1.1　常见执行机构

在这里我们将执行机构通常定义为动力原件＋相关机械结构。

常见的动力原件包括：电机、液压装置、气压装置等，当前大多数机器人中会选择电机作为动力原件。

对于移动机器人来说，常见的运动底盘有：两轮差动底盘、三轮全向底盘、四轮滑移底盘、履带式底盘、阿克曼机构运动底盘、足式运动结构等，如图 9.4 所示。

阿克曼结构底盘

四轮滑移底盘

两轮差动底盘

履带式运动底盘

足式运动结构

图 9.4　常见机器人运动机构

bobac 教育机器人运动结构为两轮差动,执行结构为两个直流电机。

9.1.2 驱动系统实现

在机器人系统中电机的控制是最常见的驱动系统,电机驱动一般通过编码器反馈实现电机的电流、速度、位置的控制。

如图 9.5 所示为 bobac 机器人驱动系统实现,boabc 驱动系统采用 STM32 单片机为主控芯片。主要提供以下功能服务:

① 电机驱动　包含四路电机驱动以及编码器采集接口;

② 电源管理　提供 12 V、5 V 等电源接口;

③ 基础传感器采集　防撞、防跌、超声波、温湿度、烟雾、电池电压监测、其他接口预留。

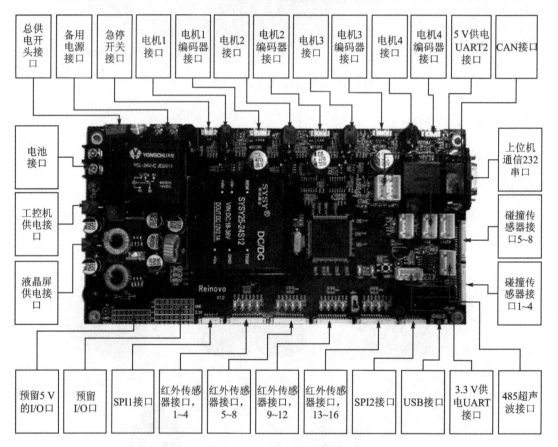

图 9.5　bobac 机器人驱动系统实现

9.1.3 传感器系统

机器人传感系统可分为外部传感器和内部传感器。

内部传感器是指测量机器人自身姿态和状态的传感器。常见的机器人内部传感器有以下几种:

编码器:通常安装在电机转动轴上用于测量电机的转速以及转动角度等,以便获取机器人的姿态。

电机编码器的工作原理如图 9.6 所示。

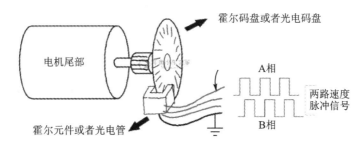

图 9.6　电机编码器工作原理

以移动机器人为例编码器的使用方法如下：

① 根据单位时间内产生的脉冲数计算电机/轮子的旋转圈数；

② 根据轮子的周长计算机器人的运动速度；

③ 根据运动速度积分计算里程。

惯性测量单元 IMU 主要用于测量机器人的姿态，如图 9.7 所示，可以测量机器人在 x,y,z 方向上的加速度，绕 x,y,z 三个轴的角速度以及 x,y,z 方向的机场大小。

外部传感器帮助机器人采集环境数据通常包含：RGB 相机、RGBD 相机、二维激光雷达、三维激光雷达、GPS 等，如图 9.8 所示。

图 9.7　IMU 传感器

　　RGB相机　　　　　　　RGBD相机　　　　　　　　GPS

　　超声波　　　　　　二维激光雷达　　　　　　三维激光雷达

图 9.8　采集环境数据的工具

9.1.4 控制系统实现

如图 9.9 所示控制器处理传感器采集的数据处理后产生合理的决策控制执行机构运动。

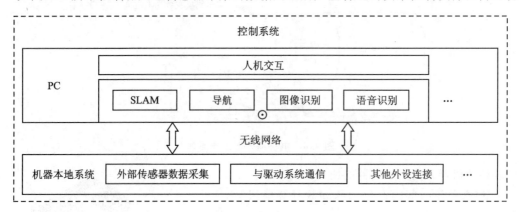

图 9.9 常用控制系统实现框架

常用机器人控制器如图 9.10 所示。

图 9.10 常用机器人控制器

9.2 URDF 描述语言解释

在很多应用和算法中,我们都需要知道机器人的大小、形状以及机器人上各个部件、传感器之间的相对安装位置。ROS 提供了一种方法允许以 xml 的格式来描述机器人,这就是本节中要介绍的 URDF。

Unified Robot Description Format,统一机器人描述格式,简称为 URDF。URDF 文件使用 XML 格式描述机器人模型。URDF 中定义机器人各个部件的几何形状、物理特性以及部件之间的连接关系。

URDF 可以帮助我们建立一个虚拟机器人,也可以告诉系统真实的机器人长什么样子。

URDF 描述文件通常由以下几个描述标签组成如表 9.1 所列。

表 9.1 URDF 常用标签

名 称	描述内容
link	用于描述机器人组成部件的动力学和运动学特性
joint	用于描述关节的运动学和动力学特性
gazebo	用于描述仿真相关物理属性和传感器属性,如惯量、摩擦系数、相机特性、激光雷达特性等
robot	用于描述机器人整体的动力学和运动学特性,包含了 link、joint、gazebo 等

(1) <link> 标签

<link> 标签用于描述机器人某个刚体部分的外观和物理属性,包括尺寸(size)、颜色(color)、形状(shap)、惯性矩阵(inertial matrix)、碰撞参数(collision properties)等。

```
<link name = "link name">
    <inertial> ...... </inertial>
    <visual> ...... </visual>
    <collision> ...... </collision>
</link>
```

<visual> 标签用于描述机器人关节(link)部分的外观参数,<inertial> 标签用于描述关节的惯性参数,而 <collision> 标签用于描述关节的碰撞属性。由图 9.11 中可以看出,碰撞(collision)的区域要大于外观(visual)的区域,这就意味着只要有其他物体与碰撞区域相交,就认为关节发生碰撞。

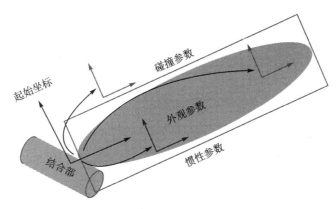

图 9.11 <link> 标签标描述

(2) <joint> 标签

<joint> 标签用于描述机器人的关节与关节之间的连接关系(见图 9.12)。
- 描述机器人关节的运动学和动力学属性;
- 包括关节运动的位置和速度限制;
- 根据关节运动形式,可以将结合部(joint)分为 6 种类型,如表 9.2 所列。

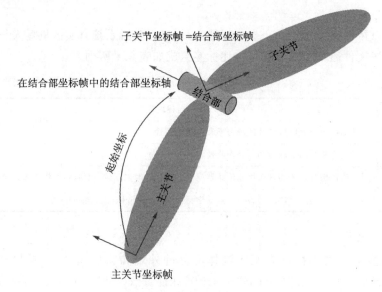

图 9.12 <joint> 标签描述

表 9.2 结合部的类型

结合部类型	描　述
continuous	旋转关节，表示该关节可以围绕某一单轴无限旋转，比如车轮和车体之间的关节
revolute	旋转关节，类似于 continous，但是旋转的角度有限制
prismatic	滑动关节，沿某一轴线移动的关节，带有位置极限
planar	平面关节，允许在平面正交方向上平移或者旋转
floating	浮动关节，允许进行平移、旋转运动
fixed	固定关节，不允许运动的特殊关节

<joint> 标签语法如下：

```
<joint name = " <name of the joint> ">
    <parent link = "parent_link"/>
    <child link = "child_link"/>
    <calibration .... />
    <dynamics damping ..... />
    <limit effort ... />
    . . . .
</joint>
```

其中，必须指定结合部（joint）的名字以及主关节（parent link）和子关节（child link），还可以设置关节的其他属性。

<calibration>：关节的参考位置，用来校准关节的绝对位置。

<dynamics>：用于描述关节的物理属性，例如阻尼值、物理静摩擦力等，经常在动力学仿真中用到。

<limit>：用于描述运动的一些极限值，包括关节运动的上下限位置、速度限制、力矩限制等。

<mimic>：用于描述该关节与已有关节的关系。
<safety_controller>：用于描述安全控制器参数。

（3）<robot> 标签

<robot> 标签是完整机器人模型的最顶层标签，<link> 和 <joint> 标签都必须包含在 <robot> 标签内。如图 9.13 所示，一个完整的机器人模型由一系列 <link> 和 <joint> 组成。

<robot> 标签内可以设置机器人的名称，其基本语法如下：

```
<robot name = " <name of the robot> ">
    <link> ....... </link>
    <link> ....... </link>
    <joint> ....... </joint>
    <joint> ....... </joint>
</robot>
```

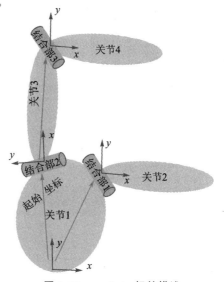

图 9.13 <robot> 标签描述

（4）<gazebo> 标签

<gazebo> 标签用于描述机器人模型在 Gazebo 中仿真所需要的参数，包括机器人材料的属性、Gazebo 插件等。该标签不是机器人模型必需的部分，只有在 gazebo 仿真时才需加入。

该标签的基本语法如下：

```
<gazebo reference = "link_1">
    <material> Gazebo/Black </material>
</gazebo>
```

本章后续内容还会通过实例继续深入讲解这些 URDF 文件中 XML 标签的使用方法。

9.3 创建 URDF 模型

在 ROS 中，机器人的模型一般放在 RobotName_description 功能包下。下面尝试从零开始创建一个移动机器人的 URDF 模型。

9.3.1 创建机器人描述功能包

在工作空间 ros_workspace 中创建一个机器人描述功能包 mbot_description。

```
cd ~/ros_workspace/src
catkin_create_pkg mbot_description urdf xacro
```

然后在工作空间中创建以下几个文件夹，如图 9.14 所示。
其中：
urdf   用于存放机器人模型的 URDF 或 xacro 文件。
meshes   用于放置 URDF 中引用的模型渲染文件。
launch   用于保存相关启动文件。
config   用于保存 rviz 的配置文件。

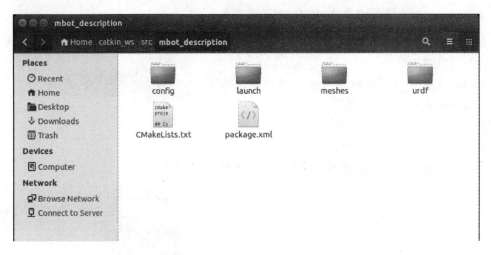

图 9.14　mbot_description 功能包

9.3.2　创建 URDF 模型

接下来我们来创建如图 9.15 所示的机器人的描述文件。

图 9.15 中,该机器人由两个转动轮、两个支撑轮加上车体共五个部件组成,另外拥有左轮到车体、右轮到车体、前支撑轮到车体、后支撑轮到车体共四个关节。

在 urdf 文件夹下创建一个文件夹 urdf,并在 urdf 文件夹下创建 mbot_base.urdf 文件,写入以下代码:

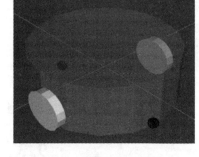

图 9.15　需要创建的机器人

```
<?xml version = "1.0" ?>
<robot name = "mbot">
    <link name = "base_link">
        <visual>
            <origin xyz = "0 0 0" rpy = "0 0 0"/>
            <geometry>
                <cylinder length = "0.16" radius = "0.20"/>
            </geometry>
            <material name = "yellow">
                <color rgba = "1 0.4 0 1"/>
            </material>
        </visual>
    </link>
    <joint name = "left_wheel_joint" type = "continuous">
        <origin xyz = "0 0.19 - 0.05" rpy = "0 0 0"/>
        <parent link = "base_link"/>
        <child link = "left_wheel_link"/>
        <axis xyz = "0 1 0"/>
    </joint>
    <link name = "left_wheel_link">
        <visual>
```

```xml
            <origin xyz = "0 0 0" rpy = "1.5707 0 0" />
            <geometry>
                <cylinder radius = "0.06" length = "0.025"/>
            </geometry>
            <material name = "white">
                <color rgba = "1 1 1 0.9"/>
            </material>
        </visual>
    </link>
    <joint name = "right_wheel_joint" type = "continuous">
        <origin xyz = "0 -0.19 -0.05" rpy = "0 0 0"/>
        <parent link = "base_link"/>
        <child link = "right_wheel_link"/>
        <axis xyz = "0 1 0"/>
    </joint>
    <link name = "right_wheel_link">
        <visual>
            <origin xyz = "0 0 0" rpy = "1.5707 0 0" />
            <geometry>
                <cylinder radius = "0.06" length = "0.025"/>
            </geometry>
            <material name = "white">
                <color rgba = "1 1 1 0.9"/>
            </material>
        </visual>
    </link>
    <joint name = "front_caster_joint" type = "continuous">
        <origin xyz = "0.18 0 -0.095" rpy = "0 0 0"/>
        <parent link = "base_link"/>
        <child link = "front_caster_link"/>
        <axis xyz = "0 1 0"/>
    </joint>
    <link name = "front_caster_link">
        <visual>
            <origin xyz = "0 0 0" rpy = "0 0 0"/>
            <geometry>
                <sphere radius = "0.015" />
            </geometry>
            <material name = "black">
                <color rgba = "0 0 0 0.95"/>
            </material>
        </visual>
    </link>
    <joint name = "back_caster_joint" type = "continuous">
        <origin xyz = "-0.18 0 -0.095" rpy = "0 0 0"/>
        <parent link = "base_link"/>
        <child link = "back_caster_link"/>
        <axis xyz = "0 1 0"/>
    </joint>
    <link name = "back_caster_link">
        <visual>
            <origin xyz = "0 0 0" rpy = "0 0 0"/>
            <geometry>
                <sphere radius = "0.015" />
```

```
            </geometry>
            <material name = "black">
                <color rgba = "0 0 0 0.95"/>
            </material>
        </visual>
    </link>
</robot>
```

代码解析：

URDF 提供了一些命令行工具，可以帮助我们检查、梳理模型文件，需要在终端中独立安装：

```
$ sudo apt-get install liburdfdom-tools
```

然后使用 check_urdf 命令对 mbot_base.urdf 文件进行检查：

```
$ check_urdf mbot_base.urdf
```

check_urdf 命令会解析 urdf 文件，并且显示解析过程中发现的错误。如果一切正常，则在终端输出图 9.16 所示的内容。

```
reinovo@reinovo-ThinkPad-E450c:~/ros_workspace/src/mbot_description/urdf/urdf$ c
heck_urdf mbot_base.urdf
robot name is: mbot
---------- Successfully Parsed XML ---------------
root Link: base_link has 4 child(ren)
    child(1):  back_caster_link
    child(2):  front_caster_link
    child(3):  left_wheel_link
    child(4):  right_wheel_link
```

图 9.16　检查 urdf 文件

还可以使用 urdf_to_graphiz 命令查看 URDF 模型的整体结构：

```
$ urdf_to_graphiz mbot_base.urdf
```

执行 urdf_to_graphiz 命令后，会在当前目录下生成一个 pdf 文件，打开该文件，可以看到模型的整体结构图（见图 9.17）。

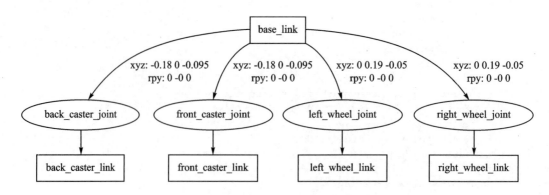

图 9.17　使用 urdf_to_graphiz 命令生成的 URDF 模型整体结构图

9.3.3 URDF 解析

针对上面创建的 URDF 模型,下面将对其关键部分进行解析。

```
<?xml version = "1.0" ?>
<robot name = "mbot_base">
```

首先需要声明该文件使用 XML 描述,然后使用 <robot> 根标签定义一个机器人模型,并定义该机器人模型的名称是 mbot_base。根标签的内容即为对机器人模型的详细定义。

```
<link name = "base_link">
    <visual>
        <origin xyz = "0 0 0" rpy = "0 0 0"/>
        <geometry>
            <cylinder length = "0.16" radius = "0.20"/>
        </geometry>
        <material name = "yellow">
            <color rgba = "1 0.4 0 1"/>
        </material>
    </visual>
</link>
```

这段代码用来描述机器人的底盘部件,<visual> 标签用来定义底盘的外观属性,在显示和仿真中,rviz 或 gazebo 会按照这里的描述将机器人模型呈现出来。我们先将机器人底盘抽象成一个圆柱结构,使用 <cylinder> 标签定义这个圆柱的半径和高;然后声明这个底盘圆柱在空间内的三维坐标位置和旋转姿态,底盘中心位于界面的中心点,所以使用 <origin> 设置起点坐标为界面的中心坐标。此外,使用 <material> 标签设置机器人底盘的颜色——黄色,其中的 <color> 便是黄色的 RGBA 值。

```
<joint name = "left_wheel_joint" type = "continuous">
    <origin xyz = "0 0.19 - 0.05" rpy = "0 0 0"/>
    <parent link = "base_link"/>
    <child link = "left_wheel_link"/>
    <axis xyz = "0 1 0"/>
</joint>
```

这段代码定义了第一个关节结合部,用来连接机器人底盘和左边轮子。结合部的类型是 continuous 类型,这种类型的结合部允许关节发生 360°运动。<origin> 标签定义了结合部的起点,我们将起点设置在需要安装轮子的位置。<rpy> 标签定义了子节点 left_wheel_link 和父节点 base_link 之间的坐标旋转关系。

```
<link name = "left_wheel_link">
    <visual>
        <origin xyz = "0 0 0" rpy = "1.5707 0 0" />
        <geometry>
            <cylinder radius = "0.06" length = "0.025"/>
        </geometry>
        <material name = "white">
            <color rgba = "1 1 1 0.9"/>
        </material>
    </visual>
```

```
</link>
```

这段代码用于描述机器人左轮部件,与机器人底盘的描述类似,<visual> 用于描述轮子的几何形状、尺寸及颜色等属性。其中 <origin xyz = "0 0 0" rpy = "1.5707 0 0" /> 表示该节点在自身坐标系中的平移和旋转在本例中表示左轮部件绕自身坐标系的 x 轴旋转 1.570 7 弧度,也就是将轮子立起来。

其他节点以及关节与以上描述雷同在此不再赘述,建议自己动手创建一个机器人描述以便加深印象。

9.3.4 在 rviz 中显示机器人模型

完成 URDF 模型的设计后,可以使用 rviz 将该模型可视化显示出来,检查是否符合设计目标。

mbot_description 功能包中创建文件夹 launch,并在 launch 文件夹下创建 urdf 文件夹,在 urdf 文件夹下创建 display_mbot_urdf.launch 文件。

在 launch 文件中写入以下代码:

```
<?xml version = "1.0"?>
<launch>
    <param name = "robot_description" textfile = "$(find mbot_description)/urdf/urdf/mbot_base.urdf" />
    <!-- 设置 GUI 参数,显示关节控制插件 -->
    <param name = "use_gui" value = "true"/>

    <!-- 运行 joint_state_publisher 节点,发布机器人的关节状态 -->
    <node name = "joint_state_publisher" pkg = "joint_state_publisher" type = "joint_state_publisher" />
    <!-- 运行 robot_state_publisher 节点,发布 tf -->
    <node name = "robot_state_publisher" pkg = "robot_state_publisher" type = "state_publisher" />
    <!-- 运行 rviz 可视化界面 -->
    <node name = "rviz" pkg = "rviz" type = "rviz" args = "-d $(find mbot_description)/config/mbot_urdf.rviz" required = "true" />
</launch>
```

代码分析:
关于 launch 文件前面已经介绍过它的语法等内容,在这里主要介绍其中的几个参数及节点。

```
<param name = "robot_description" textfile = "$(find mbot_description)/urdf/urdf/mbot_base.urdf" />
```

其中,robot_description 参数用于读取和记录 urdf 描述文件。

```
<param name = "use_gui" value = "true"/>
```

其中,use_gui 用于表征是否打开关节的控制插件。

```
<node name = "joint_state_publisher" pkg = "joint_state_publisher" type = "joint_state_publisher" />
```

该节点用于发布机器人关节状态,是运行每个机器人描述文件时都要打开的。

```
<node name = "robot_state_publisher" pkg = "robot_state_publisher" type = "state_publisher" />
```

该节点用于发布机器人各个部件之间的 tf 关系。该节点也是必须要运行的,只有该节点运行时系统才会正常发布 tf 树。

```
<node name = "rviz" pkg = "rviz" type = "rviz" args = "-d $(find mbot_description)/config/mbot_urdf.rviz" required = "true" />
```

该节点用于打开 rviz,其中的变量 args 用于加载预先配置好的 rviz 设置文件。**注意**:此时功能包中没有该变量读取的路径及文件,但是不影响我们打开 rviz,稍后我们会介绍如何保存 rviz 的配置文件。

运行 launch 文件:

```
$ roslaunch mbot_description display_mbot_urdf.launch
```

正确运行如图 9.18 所示。

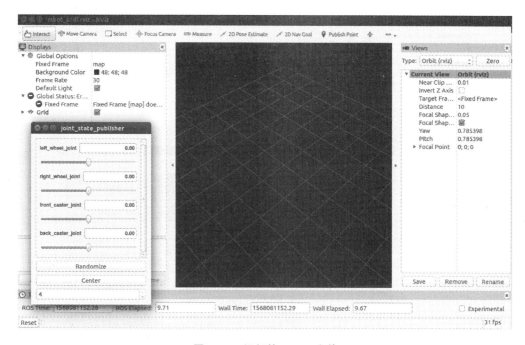

图 9.18 运行的 launch 文件

图 9.18 所示是一个空的 rviz 软件。接下来配置 rviz 软件让它显示出我们构建的机器人模型。

① 将 Fixed Frame 修改为 base_link,如图 9.19 所示,Fixed Frame 用于指定 tf 树中哪个坐标系为固定坐标。

② 添加机器人模型如图 9.20 所示。

配置完成如图 9.21 所示。

③ 保存 rviz 配置文件,在 mbot_description 功能包目录下新建 config 文件

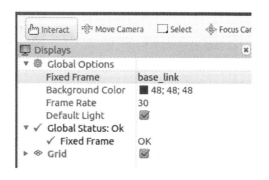

图 9.19 修改 Fixed Frame

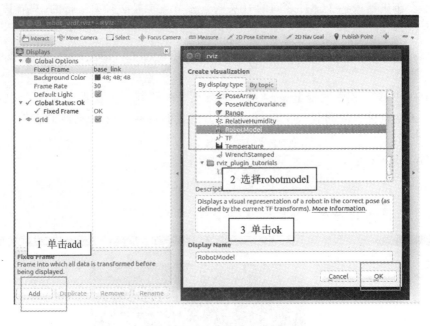

图 9.20 添加机器人模型

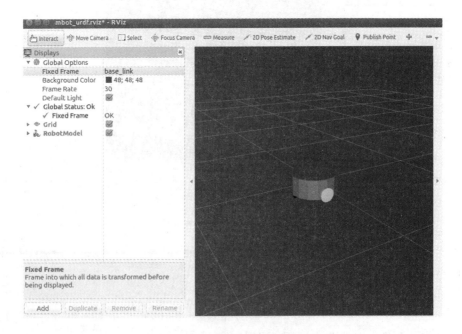

图 9.21 rviz 正常显示机器人模型

夹。单击菜单栏中 Files→save Config As 选项,如图 9.22 所示,将保存目录选择到新建的 config 文件夹中,并将配置文件命名为 mbot_urdf.rviz。

这样下次打开 rviz 时就可以直接加载这个配置文件,正常显示机器人了。

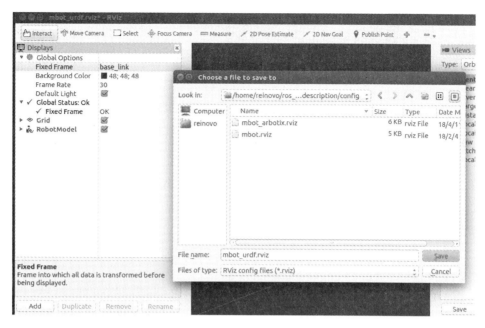

图 9.22　保存 rviz 配置文件

9.4　改进 URDF 模型

9.4.1　使用 xacro 优化 URDF 模型

回顾现在的机器人模型，我们似乎创建了一个十分冗长的模型文件，其中有很多内容除了参数，几乎都是重复的。但是 URDF 文件并不支持代码复用的特性，如果为一个复杂的机器人建模，那么 URDF 文件该有多复杂啊！

ROS 当然不会容忍这种冗长且重复的情况，因为它的设计目标就是提高代码的复用率。于是，针对 URDF 模型产生了另一种精简化、可复用、模块化的描述形式——xacro，它具备以下优势：

① 精简模型代码　xacro 是一个精简版本的 URDF 文件，在 xacro 文件中，可以通过创建宏定义的方式定义常量或者复用代码，不仅可以减少代码量，而且可以让模型代码更加模块化，更具可读性。

② 提供可编程接口　xacro 的语法支持一些可编程接口，如常量、变量、数学公式、条件语句等，可以让建模过程更加智能有效。

xacro 是 URDF 的升级版，模型文件的后缀名由 .urdf 变为 .xacro，而且在模型 <robot> 标签中需要加入 xacro 的声明：

```
<?xml version = "1.0"?>
<robot name = "robot_name" xmlns:xacro = "http://www.ros.org/wiki/xacro" >
```

1. 使用常量定义

在之前的 URDF 模型中使用了很多尺寸、坐标等常量，这些常量分布在整个文件中，不仅

可读性差,而且后期修改十分困难。xacro 提供了一种常量属性的定义方式:

```
<xacro:property name = "M_PI" value = "3.14159"/>
```

当需要使用该常量时,使用如下语法调用即可:

```
<origin xyz = "0 0 0" rpy = " $ {M_PI/2} 0 0" />
```

现在,各种参数的定义都可以使用常量定义的方式进行声明,如 9.3 节中 mbot_base 中的一些常量如下:

```
<!-- PROPERTY LIST -->
    <xacro:property name = "M_PI" value = "3.1415926"/>
    <xacro:property name = "base_radius" value = "0.20"/>
    <xacro:property name = "base_length" value = "0.16"/>

    <xacro:property name = "wheel_radius" value = "0.06"/>
    <xacro:property name = "wheel_length" value = "0.025"/>
    <xacro:property name = "wheel_joint_y" value = "0.19"/>
    <xacro:property name = "wheel_joint_z" value = "0.05"/>

    <xacro:property name = "caster_radius" value = "0.015"/>
<xacro:property name = "caster_joint_x" value = "0.18"/>
```

如果改动机器人模型,只要修改这些参数即可,十分方便。

2. 调用数学公式

在"$ {}"语句中,不仅可以调用常量,还可以使用一些常用的数学运算,包括加、减、乘、除、负号、括号等,例如:

```
<origin xyz = "0 $ {(motor_length + wheel_length)/2} 0" rpy = "0 0 0"/>
```

所有数学运算都会转换成浮点数进行,以保证运算精度。

3. 使用宏定义

xacro 文件可以使用宏定义来声明重复使用的代码模块,而且可以包含输入参数,类似编程中的函数概念。例如,mbot 中有左右两个主动轮、前后两个支撑轮,两个主动轮和支撑轮除了位置不一样,其他都一样。在 xacro 中,这种相同的模型就可以通过定义一种宏定义模块的方式来重复使用,例如下面就是主动轮的宏定义:

```
<xacro:macro name = "wheel" params = "prefix reflect">
    <joint name = " $ {prefix}_wheel_joint" type = "continuous">
        <origin xyz = "0 $ {reflect * wheel_joint_y} $ { - wheel_joint_z}" rpy = "0 0 0"/>
        <parent link = "base_link"/>
        <child link = " $ {prefix}_wheel_link"/>
        <axis xyz = "0 1 0"/>
    </joint>

    <link name = " $ {prefix}_wheel_link">
        <visual>
            <origin xyz = "0 0 0" rpy = " $ {M_PI/2} 0 0" />
            <geometry>
                <cylinder radius = " $ {wheel_radius}" length = " $ {wheel_length}"/>
            </geometry>
```

```xml
            <material name = "gray" />
        </visual>
    </link>
</xacro:macro>
```

以上宏定义中包含两个输入参数:轮子的名字以及左右的偏移。需要该宏模块时,使用如下语句调用,设置输入参数即可:

```xml
<wheel prefix = "left" reflect = " - 1"/>
```

在 mbot_description/urdf 文件夹下创建一个文件夹 xacro,并在该文件夹下创建 mbot_base.xacro 文件,输入以下代码:

```xml
<?xml version = "1.0"?>
<robot name = "mbot" xmlns:xacro = "http://www.ros.org/wiki/xacro">
    <!-- PROPERTY LIST -->
    <xacro:property name = "M_PI" value = "3.1415926"/>
    <xacro:property name = "base_radius" value = "0.20"/>
    <xacro:property name = "base_length" value = "0.16"/>

    <xacro:property name = "wheel_radius" value = "0.06"/>
    <xacro:property name = "wheel_length" value = "0.025"/>
    <xacro:property name = "wheel_joint_y" value = "0.19"/>
    <xacro:property name = "wheel_joint_z" value = "0.05"/>

    <xacro:property name = "caster_radius" value = "0.015"/>   <!-- wheel_radius - ( base_length/2 - wheel_joint_z) -->
    <xacro:property name = "caster_joint_x" value = "0.18"/>

    <!-- Defining the colors used in this robot -->
    <material name = "yellow">
        <color rgba = "1 0.4 0 1"/>
    </material>
    <material name = "black">
        <color rgba = "0 0 0 0.95"/>
    </material>
    <material name = "gray">
        <color rgba = "0.75 0.75 0.75 1"/>
    </material>

    <!-- Macro for robot wheel -->
    <xacro:macro name = "wheel" params = "prefix reflect">
        <joint name = "${prefix}_wheel_joint" type = "continuous">
            <origin xyz = "0 ${reflect * wheel_joint_y} ${- wheel_joint_z}" rpy = "0 0 0"/>
            <parent link = "base_link"/>
            <child link = "${prefix}_wheel_link"/>
            <axis xyz = "0 1 0"/>
        </joint>

        <link name = "${prefix}_wheel_link">
            <visual>
                <origin xyz = "0 0 0" rpy = "${M_PI/2} 0 0" />
                <geometry>
                    <cylinder radius = "${wheel_radius}" length = "${wheel_length}"/>
```

```xml
                </geometry>
                <material name = "gray" />
            </visual>
        </link>
    </xacro:macro>

    <!-- Macro for robot caster -->
    <xacro:macro name = "caster" params = "prefix reflect">
        <joint name = "${prefix}_caster_joint" type = "continuous">
            <origin xyz = "${reflect * caster_joint_x} 0 ${-(base_length/2 + caster_radius)}" rpy = "0 0 0"/>
            <parent link = "base_link"/>
            <child link = "${prefix}_caster_link"/>
            <axis xyz = "0 1 0"/>
        </joint>

        <link name = "${prefix}_caster_link">
            <visual>
                <origin xyz = "0 0 0" rpy = "0 0 0"/>
                <geometry>
                    <sphere radius = "${caster_radius}" />
                </geometry>
                <material name = "black" />
            </visual>
        </link>
    </xacro:macro>

    <xacro:macro name = "mbot_base">
        <link name = "base_footprint">
            <visual>
                <origin xyz = "0 0 0" rpy = "0 0 0" />
                <geometry>
                    <box size = "0.001 0.001 0.001" />
                </geometry>
            </visual>
        </link>

        <joint name = "base_footprint_joint" type = "fixed">
            <origin xyz = "0 0 ${base_length/2 + caster_radius * 2}" rpy = "0 0 0" />
            <parent link = "base_footprint"/>
            <child link = "base_link" />
        </joint>

        <link name = "base_link">
            <visual>
                <origin xyz = "0 0 0" rpy = "0 0 0" />
                <geometry>
                    <cylinder length = "${base_length}" radius = "${base_radius}"/>
                </geometry>
                <material name = "yellow" />
            </visual>
        </link>

        <wheel prefix = "left" reflect = "-1"/>
```

```
            <wheel prefix = "right" reflect = "1"/>

            <caster prefix = "front" reflect = " - 1"/>
            <caster prefix = "back" reflect = "1"/>
    </xacro:macro>
</robot>
```

9.4.2 引用 xacro 文件

在 mbot_description/urdf/xacro 文件下创建 mbot.xacro 文件,写入以下代码:

```
<?xml version = "1.0"?>
<robot name = "arm" xmlns:xacro = "http://www.ros.org/wiki/xacro">
    <xacro:include filename = "$(find mbot_description)/urdf/xacro/mbot_base.xacro" />
    <mbot_base/>
</robot>
```

<robot> 标签之间只有两行代码:

`<xacro:include filename = "$(find mbot_description)/urdf/xacro/mbot_base.xacro" />`

这行代码描述该 xacro 文件所包含的其他 xacro 文件,类似于 C 语言中的 include 文件。声明包含关系后,该文件就可以使用被包含文件中的模块了。

`<mbot_base/>`

这行代码就调用了被包含文件 mbot_base.xacro 中的机器人模型宏定义。也就是说,机器人的模型文件全部是在 mbot_base.xacro 中使用一个宏来描述的,那么为什么还需要 mbot.xacro 来包含调用呢?这是因为我们把机器人本体看作一个模块,如果需要与其他模块集成,使用这种方法就不需要修改机器人的模型文件,只需要在上层实现一个拼装模块的顶层文件即可,灵活性更强。比如后续在机器人模型上装配 camera、Kinect、rplidar,只需要修改这里的 mbot.xacro 即可。

9.4.3 显示优化后的模型

xacro 文件设计完成后,可以通过两种方式将优化后的模型显示在 rviz 中:

1. 将 xacro 文件转换成 URDF 文件

使用如下命令可以将 xacro 文件转换成 URDF 文件:

```
$ rosrun xacro xacro.py mbot.xacro > mbot.urdf
```

当前目录下会生成一个转化后的 URDF 文件,然后使用上面介绍的 launch 文件可将该 URDF 模型显示在 rviz 中。

2. 直接调用 xacro 文件解析器

也可以省略手动转换模型的过程,直接在启动文件中调用 xacro 解析器,自动将 xacro 转换成 URDF 文件。该过程可以在 launch 文件中使用如下语句进行配置:

```
<arg name = "model" default = "$(find xacro)/xacro -- inorder '$(find mbot_description)/urdf/xacro/mbot.xacro" />
<param name = "robot_description" command = "$(arg model)" />
```

在 mbot_description/launch 目录下新建一个 xacro 文件夹,并在该文件夹下新建 display_mbot_xacro.launch 文件,打开该文件输入以下代码:

```
<launch>
    <arg name = "model" default = " $ (find xacro)/xacro -- inorder '$ (find mbot_description)/urdf/xacro/mbot.xacro'" />
    <arg name = "gui" default = "true" />

    <param name = "robot_description" command = " $ (arg model)" />

    <!-- 设置 GUI 参数,显示关节控制插件 -->
    <param name = "use_gui" value = " $ (arg gui)"/>

    <!-- 运行 joint_state_publisher 节点,发布机器人的关节状态 -->
    <node name = "joint_state_publisher" pkg = "joint_state_publisher" type = "joint_state_publisher" />

    <!-- 运行 robot_state_publisher 节点,发布 tf -->
    <node name = "robot_state_publisher" pkg = "robot_state_publisher" type = "robot_state_publisher" />

    <!-- 运行 rviz 可视化界面 -->
    <node name = "rviz" pkg = "rviz" type = "rviz" args = " - d $ (find mbot_description)/config/mbot.rviz" required = "true" />

</launch>
```

launch 文件的内容在前面已经介绍过了,在此不再赘述,运行上述 launch 文件,结果如图 9.23 所示。

```
$ roslaunch mbot_description display_mbot_xacro.launch
```

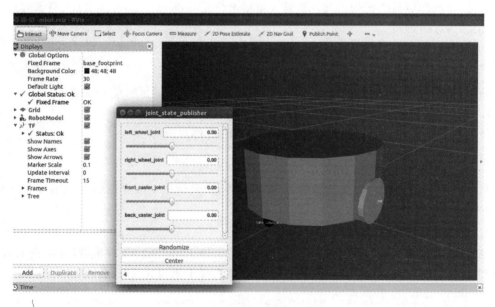

图 9.23 运行 launch 文件结果

9.5 添加传感器

在 9.3 节和 9.4 节中我们构建了一个机器人底盘。通常室内移动机器人会装配彩色摄像头、RGB-D 摄像头、激光雷达等传感器,也许现实中我们无法拥有这些传感器,但是在虚拟的机器人模型世界里我们可以创造一切。

9.5.1 添加摄像头

首先尝试创建一个摄像头的模型。笔者仿照真实摄像头画了一个长方体,以此代表摄像头模型。在 mbot_description/urdf/xacro 路径下创建一个 sensors 的文件夹,并在文件夹下创建 camera.xacro 文件。输入以下代码:

```xml
<?xml version = "1.0"?>
<robot xmlns:xacro = "http://www.ros.org/wiki/xacro" name = "camera">

    <xacro:macro name = "usb_camera" params = "prefix: = camera">
        <link name = "${prefix}_link">
            <visual>
                <origin xyz = "0 0 0" rpy = "0 0 0" />
                <geometry>
                    <box size = "0.01 0.04 0.04" />
                </geometry>
                <material name = "black"/>
            </visual>
        </link>
    </xacro:macro>

</robot>
```

上述代码中使用了一个名为 usb_camera 的宏来描述摄像头,输入参数是摄像头的名称,宏中包含了表示摄像头长方体部件的参数。

修改路径 mbot_description/urdf/xacro/下的 mbot.xacro 文件,代码如下:

```xml
<?xml version = "1.0"?>
<robot name = "arm" xmlns:xacro = "http://www.ros.org/wiki/xacro">

    <xacro:include filename = "$(find mbot_description)/urdf/xacro/mbot_base.xacro" />
    <xacro:include filename = "$(find mbot_description)/urdf/xacro/sensors/camera.xacro" />

    <xacro:property name = "camera_offset_x" value = "0.17" />
    <xacro:property name = "camera_offset_y" value = "0" />
    <xacro:property name = "camera_offset_z" value = "0.10" />
    <mbot_base/>
    <!-- Camera -->
    <joint name = "camera_joint" type = "fixed">
        <origin xyz = "${camera_offset_x} ${camera_offset_y} ${camera_offset_z}" rpy = "0 0 0" />
        <parent link = "base_link" />
        <child link = "camera_link" />
    </joint>
    <xacro:usb_camera prefix = "camera" />
</robot>
```

在这个顶层 xacro 文件中,包含了描述摄像头以及底盘的模型文件,然后使用一个 fixed 类型的结合部把摄像头固定到机器人顶部支撑板靠前的位置。

运行如下命令在 rviz 中查看生成的机器人模型,如图 9.24 所示:

```
$ roslaunch mbot_description display_mbot_xacro.launch
```

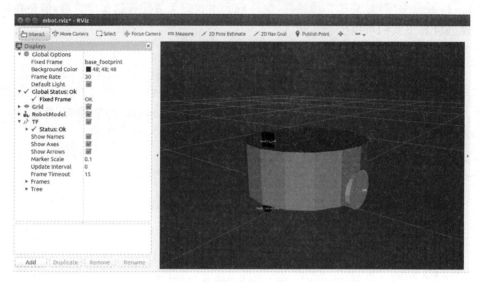

图 9.24 机器人添加 USB 摄像头

9.5.2 添加激光雷达

激光雷达的添加与摄像头的添加类似,首先在路径 mbot_description/urdf/xacro/sensors 下创建文件 lidar.xacro,写入以下代码:

```
<?xml version = "1.0"?>
<robot xmlns:xacro = "http://www.ros.org/wiki/xacro" name = "laser">

    <xacro:macro name = "rplidar" params = "prefix: = laser">
        <link name = " $ {prefix}_link">
            <visual>
                <origin xyz = " 0 0 0 " rpy = "0 0 0 " />
                <geometry>
                    <cylinder length = "0.05" radius = "0.05"/>
                </geometry>
                <material name = "black"/>
            </visual>
        </link>
    </xacro:macro>

</robot>
```

在该描述文件中定义了雷达的宏,该宏拥有一个参数 prefix,该参数指明了雷达节点的名字。

修改 mbot_description/urdf/xacro 路径下的 mbot.xacro,将激光雷达安装到机器人上,代码如下:

```xml
<?xml version = "1.0"?>
<robot name = "arm" xmlns:xacro = "http://www.ros.org/wiki/xacro">

    <xacro:include filename = "$(find mbot_description)/urdf/xacro/mbot_base.xacro" />
    <xacro:include filename = "$(find mbot_description)/urdf/xacro/sensors/camera.xacro" />
    <xacro:include filename = "$(find mbot_description)/urdf/xacro/sensors/lidar.xacro" />

    <xacro:property name = "camera_offset_x" value = "0.17" />
    <xacro:property name = "camera_offset_y" value = "0" />
    <xacro:property name = "camera_offset_z" value = "0.10" />

    <xacro:property name = "lidar_offset_x" value = "0" />
    <xacro:property name = "lidar_offset_y" value = "0" />
    <xacro:property name = "lidar_offset_z" value = "0.105" />

    <mbot_base/>

    <!-- 摄像头 -->
    <joint name = "camera_joint" type = "fixed">
        <origin xyz = "${camera_offset_x} ${camera_offset_y} ${camera_offset_z}" rpy = "0 0 0" />
        <parent link = "base_link"/>
        <child link = "camera_link"/>
    </joint>

    <joint name = "lidar_joint" type = "fixed">
        <origin xyz = "${lidar_offset_x} ${lidar_offset_y} ${lidar_offset_z}" rpy = "0 0 0" />
        <parent link = "base_link"/>
        <child link = "laser_link"/>
    </joint>

    <xacro:usb_camera prefix = "camera"/>
    <xacro:rplidar prefix = "laser"/>

</robot>
```

运行 launch 文件查看机器人模型,如图 9.25 所示。

```
$ roslaunch mbot_description display_mbot_xacro.launch
```

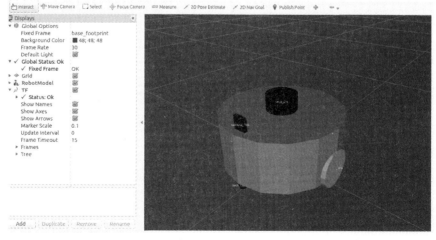

图 9.25　给机器人安装激光雷达

如果需要添加其他传感器，按照上面案例操作即可。

9.6 gazebo 仿真

前面介绍了如何创建一个机器人模型，但是该模型只是描述了机器人以及传感器的外形，如果让机器人能够拥有真实的物理属性，传感器可以像真实传感器一样输出数据，那么需要给机器人添加更多的描述：

① 给每个 <link> 添加惯性属性、碰撞属性及 gazebo 属性；
② 给机器人添加控制器；
③ 给传感器添加传感器插件。

9.6.1 给 base_link 添加惯性、碰撞及 gazebo 属性

在路径 mbot_description/urdf/xacro 文件夹下新建文件夹 gazebo，并在 gazebo 文件下创建 mbot_base_gazebo.xacro 文件。代码如下：

```xml
<?xml version = "1.0"?>
<robot name = "mbot" xmlns:xacro = "http://www.ros.org/wiki/xacro">
    <!-- 常量列表 -->
    <xacro:property name = "M_PI" value = "3.1415926"/>
    <xacro:property name = "base_mass"   value = "20" />
    <xacro:property name = "base_radius" value = "0.20"/>
    <xacro:property name = "base_length" value = "0.16"/>

    <xacro:property name = "wheel_mass"   value = "2" />
    <xacro:property name = "wheel_radius" value = "0.06"/>
    <xacro:property name = "wheel_length" value = "0.025"/>
    <xacro:property name = "wheel_joint_y" value = "0.19"/>
    <xacro:property name = "wheel_joint_z" value = "0.05"/>

    <xacro:property name = "caster_mass"    value = "0.5" />
    <xacro:property name = "caster_radius"  value = "0.015"/>
    <!-- wheel_radius - ( base_length/2 - wheel_joint_z) -->
    <xacro:property name = "caster_joint_x" value = "0.18"/>

    <!-- 颜色定义 -->
    <material name = "yellow">
        <color rgba = "1 0.4 0 1"/>
    </material>
    <material name = "black">
        <color rgba = "0 0 0 0.95"/>
    </material>
    <material name = "gray">
        <color rgba = "0.75 0.75 0.75 1"/>
    </material>
    <!-- 球形转动惯量矩阵宏定义 -->
    <xacro:macro name = "sphere_inertial_matrix" params = "m r">
        <inertial>
            <mass value = "${m}" />
            <inertia ixx = "${2*m*r*r/5}" ixy = "0" ixz = "0"
```

```xml
            iyy = " $ {2 * m * r * r/5}" iyz = "0"
            izz = " $ {2 * m * r * r/5}" />
        </inertial>
    </xacro:macro>
    <!-- 圆柱体转动惯量矩阵宏定义 -->
    <xacro:macro name = "cylinder_inertial_matrix" params = "m r h">
        <inertial>
            <mass value = " $ {m}" />
            <inertia ixx = " $ {m * (3 * r * r + h * h)/12}" ixy = "0" ixz = "0"
                iyy = " $ {m * (3 * r * r + h * h)/12}" iyz = "0"
                izz = " $ {m * r * r/2}" />
        </inertial>
    </xacro:macro>
    <!-- 车轮宏定义 -->
    <xacro:macro name = "wheel" params = "prefix reflect">
        <joint name = " $ {prefix}_wheel_joint" type = "continuous">
            <origin xyz = "0 $ {reflect * wheel_joint_y} $ { - wheel_joint_z}" rpy = "0 0 0"/>
            <parent link = "base_link"/>
            <child link = " $ {prefix}_wheel_link"/>
            <axis xyz = "0 1 0"/>
        </joint>

        <link name = " $ {prefix}_wheel_link">
            <visual>
                <origin xyz = "0 0 0" rpy = " $ {M_PI/2} 0 0" />
                <geometry>
                    <cylinder radius = " $ {wheel_radius}" length = " $ {wheel_length}"/>
                </geometry>
                <material name = "gray" />
            </visual>
            <collision>
                <origin xyz = "0 0 0" rpy = " $ {M_PI/2} 0 0" />
                <geometry>
                    <cylinder radius = " $ {wheel_radius}" length = " $ {wheel_length}"/>
                </geometry>
            </collision>
            <cylinder_inertial_matrix  m = " $ {wheel_mass}" r = " $ {wheel_radius}" h = " $ {wheel_length}" />
        </link>

        <gazebo reference = " $ {prefix}_wheel_link">
            <material> Gazebo/Gray </material>
        </gazebo>
    </xacro:macro>

    <!-- 支撑轮宏定义 -->
    <xacro:macro name = "caster" params = "prefix reflect">
        <joint name = " $ {prefix}_caster_joint" type = "continuous">
            <origin xyz = " $ {reflect * caster_joint_x} 0 $ { - (base_length/2 + caster_radius)}" rpy = "0 0 0"/>
            <parent link = "base_link"/>
            <child link = " $ {prefix}_caster_link"/>
            <axis xyz = "0 1 0"/>
        </joint>
```

```xml
<link name = "${prefix}_caster_link">
    <visual>
        <origin xyz = "0 0 0" rpy = "0 0 0"/>
        <geometry>
            <sphere radius = "${caster_radius}" />
        </geometry>
        <material name = "black" />
    </visual>
    <collision>
        <origin xyz = "0 0 0" rpy = "0 0 0"/>
        <geometry>
            <sphere radius = "${caster_radius}" />
        </geometry>
    </collision>
    <sphere_inertial_matrix  m = "${caster_mass}" r = "${caster_radius}" />
</link>

<gazebo reference = "${prefix}_caster_link">
    <material> Gazebo/Black </material>
</gazebo>
</xacro:macro>
<!-- 机器人本体宏定义 -->
<xacro:macro name = "mbot_base_gazebo">
    <link name = "base_footprint">
        <visual>
            <origin xyz = "0 0 0" rpy = "0 0 0" />
            <geometry>
                <box size = "0.001 0.001 0.001" />
            </geometry>
        </visual>
    </link>
    <gazebo reference = "base_footprint">
        <turnGravityOff> false </turnGravityOff>
    </gazebo>

    <joint name = "base_footprint_joint" type = "fixed">
        <origin xyz = "0 0 ${base_length/2 + caster_radius * 2}" rpy = "0 0 0" />
        <parent link = "base_footprint"/>
        <child link = "base_link" />
    </joint>

    <link name = "base_link">
        <visual>
            <origin xyz = " 0 0 0" rpy = "0 0 0" />
            <geometry>
                <cylinder length = "${base_length}" radius = "${base_radius}"/>
            </geometry>
            <material name = "yellow" />
        </visual>
        <collision>
            <origin xyz = " 0 0 0" rpy = "0 0 0" />
            <geometry>
                <cylinder length = "${base_length}" radius = "${base_radius}"/>
            </geometry>
```

```xml
        </collision>
        <cylinder_inertial_matrix m = "${base_mass}" r = "${base_radius}" h = "${base_length}" />
    </link>

    <gazebo reference = "base_link">
        <material> Gazebo/Blue </material>
    </gazebo>

    <wheel prefix = "left"  reflect = "-1"/>
    <wheel prefix = "right" reflect = "1"/>

    <caster prefix = "front" reflect = "-1"/>
    <caster prefix = "back"  reflect = "1"/>
    </xacro:macro>
</robot>
```

1. 惯性参数

```xml
<inertial>
    <mass value = "${m}" />
    <inertia ixx = "${m*(3*r*r+h*h)/12}" ixy = "0" ixz = "0"
             iyy = "${m*(3*r*r+h*h)/12}" iyz = "0"
             izz = "${m*r*r/2}" />
</inertial>
```

其中,惯性参数的设置主要包含质量和惯性矩阵。如果是规则物体,可以通过尺寸、质量等公式计算得到惯性矩阵,你可以自行上网搜索相应的计算公式,这里使用一组虚拟的惯性矩阵数据。

2. 碰撞参数

```xml
<collision>
    <origin xyz = "0 0 0" rpy = "0 0 0" />
    <geometry>
        <cylinder length = "${base_length}" radius = "${base_radius}"/>
    </geometry>
</collision>
```

<collision>标签中的内容与<visual>标签中的内容几乎一致,这是因为我们使用的模型都是较为简单的规则模型,如果使用真实机器人的设计模型,<visual>标签中可以显示复杂的机器人外观,但是为了减少碰撞检测时的计算量,<collision>中往往使用简化后的机器人模型,例如可以将机械臂的一根连杆简化成圆柱体或长方体。

3. <gazebo>标签

针对机器人模型,需要给每一个部件添加<gazebo>标签,包含的属性仅有材质。材质属性的作用与部件里<visual>中材质属性的作用相同,gazebo无法通过<visual>中的材质参数设置外观颜色,所以需要单独设置,否则默认情况下gazebo中显示的模型全是灰白色。

```xml
<gazebo reference = "base_link">
    <material> Gazebo/Blue </material>
</gazebo>
```

9.6.2 在 gazebo 中显示机器人模型

为了便于讲解后面的内容,我们先在 mbot_description/urfd/xacro/gazebo 目录下创建 mbot_gazebo.xacro 文件,并输入以下代码:

```xml
<?xml version = "1.0"?>
<robot name = "arm" xmlns:xacro = "http://www.ros.org/wiki/xacro">

    <xacro:include filename = "$(find mbot_description)/urdf/xacro/gazebo/mbot_base_gazebo.xacro" />

    <mbot_base_gazebo/>

</robot>
```

在该文件中通过宏调用了 9.6.1 小节中创建的机器人 gazebo 模型。

在功能包 mbot_description/launch 路径下创建一个文件夹 gazebo,用于存放在 gazebo 中显示机器人模型的启动文件。在 gazebo 文件夹中新建一个 launch 文件,命名为 display_mbot_gazebo.launch,并输入以下代码:

```xml
<launch>
    <!-- 设置 launch 文件的参数 -->
    <arg name = "paused" default = "false"/>
    <arg name = "use_sim_time" default = "true"/>
    <arg name = "gui" default = "true"/>
    <arg name = "headless" default = "false"/>
    <arg name = "debug" default = "false"/>

    <!-- 运行 gazebo 仿真环境 -->
    <include file = "$(find gazebo_ros)/launch/empty_world.launch">
        <arg name = "debug" value = "$(arg debug)" />
        <arg name = "gui" value = "$(arg gui)" />
        <arg name = "paused" value = "$(arg paused)"/>
        <arg name = "use_sim_time" value = "$(arg use_sim_time)"/>
        <arg name = "headless" value = "$(arg headless)"/>
    </include>

    <!-- 加载机器人模型描述参数 -->
    <param name = "robot_description" command = "$(find xacro)/xacro -- inorder '$(find mbot_description)/urdf/xacro/gazebo/mbot_gazebo.xacro'" />

    <!-- 运行 joint_state_publisher 节点,发布机器人的关节状态 -->
    <node name = "joint_state_publisher" pkg = "joint_state_publisher" type = "joint_state_publisher" > </node>

    <!-- 运行 robot_state_publisher 节点,发布 tf -->
    <node name = "robot_state_publisher" pkg = "robot_state_publisher" type = "robot_state_publisher" output = "screen" >
        <param name = "publish_frequency" type = "double" value = "50.0" />
    </node>

    <!-- 在 gazebo 中加载机器人模型 -->
```

```
<node name = "urdf_spawner" pkg = "gazebo_ros" type = "spawn_model" respawn = "false" output = "screen" args = " - urdf  - model mbot  - param robot_description"/>
```

```
</launch>
```

运行以下命令：

```
$ roslaunch mbot_description display_mbot_gazebo.launch
```

该 launch 文件结果如图 9.26 所示。

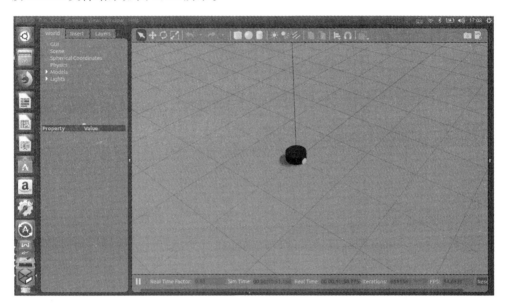

图 9.26　在 gazebo 中显示机器人模型

9.6.3　gazebo 常用插件

在 9.6.2 小节中，我们在 gazebo 中建立了机器人的三维仿真模型。该三维模型具备基本的物理属性，但是如果要模拟真实的机器人还需要控制器以及各种传感器，本小节介绍如何给仿真机器人添加控制器及传感器，使仿真机器人更加真实。

在 gazebo 中控制器及传感器一般用插件的形式来描述。

gazebo 插件可以根据插件的作用范围应用到 URDF 模型的 <robot>、<link>、<joint> 上，需要使用 <gazebo> 标签作为封装。

1. 为 <robot> 元素添加插件

为 <robot> 元素添加 gazebo 插件的方式如下：

```
<gazebo>
<plugin name = "unique_name" filename = "plugin_name.so">
    ... plugin parameters ...
    </plugin>
</gazebo>
```

与其他的 <gazebo> 元素相同，如果 <gazebo> 元素中没有设置 reference＝"x"属性，则默认应用于 <robot> 标签。

2. 为 <link>、<joint> 标签添加插件

如果需要为 <link>、<joint> 标签添加插件,则需要设置 <gazebo> 标签中的 reference = "x"属性:

```
<gazebo reference = "your_link_name">
    <plugin name = " unique_name " filename = "plugin_name.so">
        ... plugin parameters ...
    </plugin>
</gazebo>
```

至于 gazebo 目前支持的插件种类,可以查看 ROS 默认安装路径下的/opt/ros/kinetic/lib 文件夹,所有插件都是以 libgazeboXXX.so 的形式命名的。

(1) 添加控制器插件

在本案例中使用了一个用于差速控制的插件 libgazebo_ros_diff_drive.so。该插件可以接收 geometry_msgs/Twist 消息类型的速度控制指令,从而控制虚拟机器人移动,同时也可以向外部发布里程计信息。具体如下:

```
<gazebo>
    <plugin name = "differential_drive_controller"
        filename = "libgazebo_ros_diff_drive.so">
        <rosDebugLevel> Debug </rosDebugLevel>
        <publishWheelTF> false </publishWheelTF>
        <robotNamespace> / </robotNamespace>
        <publishTf> 1 </publishTf>
        <publishWheelJointState> true </publishWheelJointState>
        <alwaysOn> true </alwaysOn>
        <updateRate> 100.0 </updateRate>
        <legacyMode> true </legacyMode>
        <leftJoint> left_wheel_joint </leftJoint>
        <rightJoint> right_wheel_joint </rightJoint>
        <wheelSeparation> ${wheel_joint_y * 2} </wheelSeparation>
        <wheelDiameter> ${2 * wheel_radius} </wheelDiameter>
        <broadcastTF> true </broadcastTF>
        <wheelTorque> 30 </wheelTorque>
        <wheelAcceleration> 1.8 </wheelAcceleration>
        <commandTopic> cmd_vel </commandTopic>
        <odometryFrame> odom </odometryFrame>
        <odometryTopic> odom </odometryTopic>
        <robotBaseFrame> base_footprint </robotBaseFrame>
    </plugin>
</gazebo>
</xacro:macro>
```

在加载差速控制器插件的过程中,需要配置一系列参数,其中比较关键的参数如下:

<robotNamespace> 机器人的命名空间,插件所有数据的发布、订阅都在该命名空间下。

<leftJoint> 和 <rightJoint> 左右轮转动的关节结合部,控制器插件最终需要控制这两个结合部转动。

<wheelSeparation> 和 <wheelDiameter> 这是机器人模型的相关尺寸,在计算差速参

数时需要用到。

<wheelAcceleration>　　车轮转动的加速度。

<commandTopic>　　控制器订阅的速度控制指令,ROS 中一般都命名为 cmd_vel,生成全局命名时需要结合 <robotNamespace> 中设置的命名空间。

<odometryFrame>　　里程计数据的参考坐标系,ROS 中一般都命名为 odom。

修改 mbot_base_gazebo.xacro,添加两轮差动控制器,代码如下:

```xml
<?xml version="1.0"?>
<robot name="mbot" xmlns:xacro="http://www.ros.org/wiki/xacro">
    <!-- PROPERTY LIST -->
    <xacro:property name="M_PI" value="3.1415926"/>
    <xacro:property name="base_mass"   value="20" />
    <xacro:property name="base_radius" value="0.20"/>
    <xacro:property name="base_length" value="0.16"/>

    <xacro:property name="wheel_mass"   value="2" />
    <xacro:property name="wheel_radius" value="0.06"/>
    <xacro:property name="wheel_length" value="0.025"/>
    <xacro:property name="wheel_joint_y" value="0.19"/>
    <xacro:property name="wheel_joint_z" value="0.05"/>

    <xacro:property name="caster_mass"    value="0.5" />
    <xacro:property name="caster_radius"  value="0.015"/>
    <!-- wheel_radius - ( base_length/2 - wheel_joint_z) -->
    <xacro:property name="caster_joint_x" value="0.18"/>

    <!-- 定义机器人使用的颜色 -->
    <material name="yellow">
        <color rgba="1 0.4 0 1"/>
    </material>
    <material name="black">
        <color rgba="0 0 0 0.95"/>
    </material>
    <material name="gray">
        <color rgba="0.75 0.75 0.75 1"/>
    </material>

    <!-- 惯性矩阵宏指令 -->
    <xacro:macro name="sphere_inertial_matrix" params="m r">
        <inertial>
            <mass value="${m}" />
            <inertia ixx="${2*m*r*r/5}" ixy="0" ixz="0"
                iyy="${2*m*r*r/5}" iyz="0"
                izz="${2*m*r*r/5}" />
        </inertial>
    </xacro:macro>

    <xacro:macro name="cylinder_inertial_matrix" params="m r h">
        <inertial>
            <mass value="${m}" />
            <inertia ixx="${m*(3*r*r+h*h)/12}" ixy="0" ixz="0"
                iyy="${m*(3*r*r+h*h)/12}" iyz="0"
                izz="${m*r*r/2}" />
```

```xml
        </inertial>
    </xacro:macro>
    <!-- 机器人车轮宏指令 -->
    <xacro:macro name="wheel" params="prefix reflect">
        <joint name="${prefix}_wheel_joint" type="continuous">
            <origin xyz="0 ${reflect*wheel_joint_y} ${-wheel_joint_z}" rpy="0 0 0"/>
            <parent link="base_link"/>
            <child link="${prefix}_wheel_link"/>
            <axis xyz="0 1 0"/>
        </joint>

        <link name="${prefix}_wheel_link">
            <visual>
                <origin xyz="0 0 0" rpy="${M_PI/2} 0 0" />
                <geometry>
                    <cylinder radius="${wheel_radius}" length="${wheel_length}"/>
                </geometry>
                <material name="gray" />
            </visual>
            <collision>
                <origin xyz="0 0 0" rpy="${M_PI/2} 0 0" />
                <geometry>
                    <cylinder radius="${wheel_radius}" length="${wheel_length}"/>
                </geometry>
            </collision>
            <cylinder_inertial_matrix  m="${wheel_mass}" r="${wheel_radius}" h="${wheel_length}" />
        </link>

        <gazebo reference="${prefix}_wheel_link">
            <material>Gazebo/Gray</material>
        </gazebo>
    </xacro:macro>

    <!-- 机器人脚轮宏指令 -->
    <xacro:macro name="caster" params="prefix reflect">
        <joint name="${prefix}_caster_joint" type="continuous">
            <origin xyz="${reflect*caster_joint_x} 0 ${-(base_length/2 + caster_radius)}" rpy="0 0 0"/>
            <parent link="base_link"/>
            <child link="${prefix}_caster_link"/>
            <axis xyz="0 1 0"/>
        </joint>

        <link name="${prefix}_caster_link">
            <visual>
                <origin xyz="0 0 0" rpy="0 0 0"/>
                <geometry>
                    <sphere radius="${caster_radius}" />
                </geometry>
                <material name="black" />
            </visual>
            <collision>
                <origin xyz="0 0 0" rpy="0 0 0"/>
                <geometry>
                    <sphere radius="${caster_radius}" />
```

```xml
            </geometry>
        </collision>
        <sphere_inertial_matrix  m = " ${caster_mass}" r = " ${caster_radius}" />
    </link>

    <gazebo reference = " ${prefix}_caster_link">
        <material> Gazebo/Black </material>
    </gazebo>
</xacro:macro>

<xacro:macro name = "mbot_base_gazebo">
    <link name = "base_footprint">
        <visual>
            <origin xyz = "0 0 0" rpy = "0 0 0" />
            <geometry>
                <box size = "0.001 0.001 0.001" />
            </geometry>
        </visual>
    </link>
    <gazebo reference = "base_footprint">
        <turnGravityOff> false </turnGravityOff>
    </gazebo>

    <joint name = "base_footprint_joint" type = "fixed">
        <origin xyz = "0 0 ${base_length/2 + caster_radius * 2}" rpy = "0 0 0" />
        <parent link = "base_footprint"/>
        <child link = "base_link" />
    </joint>

    <link name = "base_link">
        <visual>
            <origin xyz = " 0 0 0" rpy = "0 0 0" />
            <geometry>
                <cylinder length = " ${base_length}" radius = " ${base_radius}"/>
            </geometry>
            <material name = "yellow" />
        </visual>
        <collision>
            <origin xyz = " 0 0 0" rpy = "0 0 0" />
            <geometry>
                <cylinder length = " ${base_length}" radius = " ${base_radius}"/>
            </geometry>
        </collision>
        <cylinder_inertial_matrix   m = " ${base_mass}" r = " ${base_radius}" h = " ${base_length}" />
    </link>

    <gazebo reference = "base_link">
        <material> Gazebo/Blue </material>
    </gazebo>

    <wheel prefix = "left"   reflect = " - 1"/>
    <wheel prefix = "right" reflect = "1"/>

    <caster prefix = "front" reflect = " - 1"/>
    <caster prefix = "back"   reflect = "1"/>
```

```xml
<!-- 控制端 -->
<gazebo>
    <plugin name = "differential_drive_controller"
            filename = "libgazebo_ros_diff_drive.so">
        <rosDebugLevel> Debug </rosDebugLevel>
        <publishWheelTF> false </publishWheelTF>
        <robotNamespace> / </robotNamespace>
        <publishTf> 1 </publishTf>
        <publishWheelJointState> true </publishWheelJointState>
        <alwaysOn> true </alwaysOn>
        <updateRate> 100.0 </updateRate>
        <legacyMode> true </legacyMode>
        <leftJoint> left_wheel_joint </leftJoint>
        <rightJoint> right_wheel_joint </rightJoint>
        <wheelSeparation> ${wheel_joint_y * 2} </wheelSeparation>
        <wheelDiameter> ${2 * wheel_radius} </wheelDiameter>
        <broadcastTF> true </broadcastTF>
        <wheelTorque> 30 </wheelTorque>
        <wheelAcceleration> 1.8 </wheelAcceleration>
        <commandTopic> cmd_vel </commandTopic>
        <odometryFrame> odom </odometryFrame>
        <odometryTopic> odom </odometryTopic>
        <robotBaseFrame> base_footprint </robotBaseFrame>
    </plugin>
</gazebo>
</xacro:macro>

</robot>
```

运行以下代码,在 gazebo 中打开机器人模型,如图 9.27 所示。

```
$ roslaunch mbot_description display_mbot_gazebo.launch
```

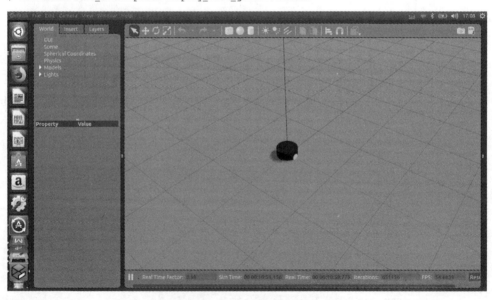

图 9.27 添加控制器后的机器人

如图 9.27 所示,添加控制器后的机器人模型与添加之前外形上并没有差别,但是此时我们可以通过发布 cmd_vel 指令来控制机器人移动。

打开一个终端,输入以下命令打开话题发布器,如图 9.28 所示:

$ rosrun rqt_publisher rqt_publisher

图 9.28 话题发布器

在 topic 下拉列表框中选择 cmd_vel 话题,并单击"+"按钮,如图 9.29 所示。

图 9.29 添加要发布的话题

可以通过修改 linear 下面的 x 值来控制机器人运动的线速度,以及 angular 下面的 z 值来控制机器人运动的角速度,如图 9.30 所示。

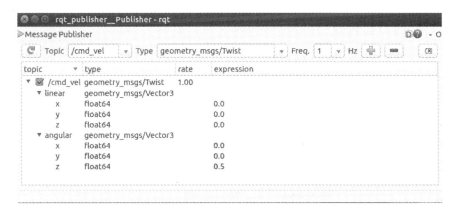

图 9.30 通过话题控制机器人运动

(2) 添加激光雷达插件

在 mbot_description/urdf/xacro/sensors 路径新建一个 xarco 文件命名为:lidar_gazebo.xacro,并输入以下代码:

```xml
<?xml version="1.0"?>
<robot xmlns:xacro="http://www.ros.org/wiki/xacro" name="laser">

    <xacro:macro name="rplidar" params="prefix:=laser">
        <!-- 创建激光雷达参考系 -->
        <link name="${prefix}_link">
            <inertial>
                <mass value="0.1" />
                <origin xyz="0 0 0" />
                <inertia ixx="0.01" ixy="0.0" ixz="0.0"
                         iyy="0.01" iyz="0.0"
                         izz="0.01" />
            </inertial>

            <visual>
                <origin xyz=" 0 0 0 " rpy="0 0 0" />
                <geometry>
                    <cylinder length="0.05" radius="0.05"/>
                </geometry>
                <material name="black"/>
            </visual>

            <collision>
                <origin xyz="0.0 0.0 0.0" rpy="0 0 0" />
                <geometry>
                    <cylinder length="0.06" radius="0.05"/>
                </geometry>
            </collision>
        </link>

        <gazebo reference="${prefix}_link">
            <material> Gazebo/Black </material>
        </gazebo>

        <gazebo reference="${prefix}_link">
            <sensor type="ray" name="rplidar">
                <pose> 0 0 0 0 0 0 </pose>
                <visualize> false </visualize>
                <update_rate> 5.5 </update_rate>
                <ray>
                    <scan>
                      <horizontal>
                        <samples> 360 </samples>
                        <resolution> 1 </resolution>
                        <min_angle> -3 </min_angle>
                        <max_angle> 3 </max_angle>
                      </horizontal>
                    </scan>
                    <range>
                      <min> 0.10 </min>
                      <max> 6.0 </max>
                      <resolution> 0.01 </resolution>
                    </range>
                    <noise>
```

```xml
            <type> gaussian </type>
            <mean> 0.0 </mean>
            <stddev> 0.01 </stddev>
          </noise>
        </ray>
        <plugin name = "gazebo_rplidar" filename = "libgazebo_ros_laser.so">
            <topicName> /scan </topicName>
            <frameName> laser_link </frameName>
        </plugin>
      </sensor>
    </gazebo>

  </xacro:macro>
</robot>
```

上面描述文件中包含了激光雷达的描述文件以及 gazebo 插件。gazebo 插件用于描述该激光雷达的参数。

```xml
<!-- reference 指定哪个 link 为激光雷达 -->
<gazebo reference = " ${prefix}_link">
            <sensor type = "ray" name = "rplidar">
                <pose> 0 0 0 0 0 0 </pose>
                <visualize> false </visualize>
                <!-- 设置更新频率 -->
                <update_rate> 5.5 </update_rate>
                <ray>
                    <scan>
                      <horizontal>
                        <!-- 采样点数量 -->
                        <samples> 360 </samples>
<!-- 采样分辨率,与上面数量相关 -->
                        <resolution> 1 </resolution>
<!-- 扫描角度范围,单位为弧度 -->
                        <min_angle> -3 </min_angle>
                        <max_angle> 3 </max_angle>
                      </horizontal>
                    </scan>
                    <range>
<!-- 扫描距离范围 -->
                        <min> 0.10 </min>
                        <max> 6.0 </max>
                        <resolution> 0.01 </resolution>
                    </range>
                    <noise>
                      <type> gaussian </type>
                      <mean> 0.0 </mean>
                      <stddev> 0.01 </stddev>
                    </noise>
                </ray>
                <plugin name = "gazebo_rplidar" filename = "libgazebo_ros_laser.so">
                    <topicName> /scan </topicName>
                    <frameName> laser_link </frameName>
                </plugin>
            </sensor>
```

```
        </gazebo>

    </xacro:macro>
</robot>
```

在 gazebo 中显示添加激光雷达后的机器人模型。首先修改 mbot_description/urdf/xacro/gazebo 下的 mbot_gazebo.xacro 文件如下：

```
<?xml version = "1.0"?>
<robot name = "arm" xmlns:xacro = "http://www.ros.org/wiki/xacro">

    <xacro:include filename = "$(find mbot_description)/urdf/xacro/gazebo/mbot_base_gazebo.xacro" />
    <xacro:include filename = "$(find mbot_description)/urdf/xacro/sensors/lidar_gazebo.xacro" />

    <xacro:property name = "lidar_offset_x" value = "0" />
    <xacro:property name = "lidar_offset_y" value = "0" />
    <xacro:property name = "lidar_offset_z" value = "0.105" />

    <!-- mbot_base /-->

    <!-- 激光雷达 -->
    <joint name = "lidar_joint" type = "fixed">
        <origin xyz = "${lidar_offset_x} ${lidar_offset_y} ${lidar_offset_z}" rpy = "0 0 0" />
        <parent link = "base_link"/>
        <child link = "laser_link"/>
    </joint>

    <xacro:rplidar prefix = "laser"/>

    <mbot_base_gazebo/>

</robot>
```

运行以下命令打开 gazebo：

```
$ roslaunch mbot_description display_mbot_gazebo.launch
```

正常运行如图 9.31 所示。

为了能够展示激光雷达数据，在 gazebo 环境中随便放入几个其他物品，在 insert 目录中选取要添加的物体，如图 9.32 所示放置了两个柜子（cabinet）。

在 rviz 中显示激光雷达的数据，修改 mbot_description/launch/xacro 路径下的 display_mbot_xacro.launch 文件，输入如下代码：

```
<launch>
    <arg name = "model" default = "$(find xacro)/xacro --inorder '$(find mbot_description)/urdf/xacro/gazebo/mbot_gazebo.xacro'" />
    <arg name = "gui" default = "true" />

    <param name = "robot_description" command = "$(arg model)" />

    <!-- 设置 GUI 参数，显示关节控制插件 -->
```

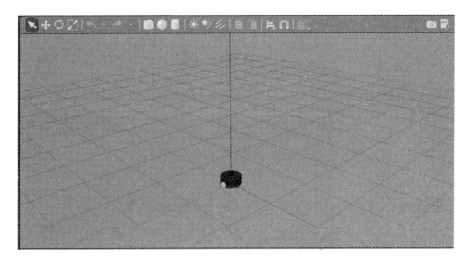

图 9.31 带激光雷达的机器人

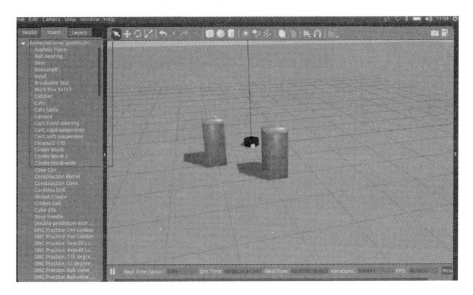

图 9.32 在 gazebo 中添加障碍

```
<param name = "use_gui" value = "$(arg gui)"/>

<!-- 运行 joint_state_publisher 节点,发布机器人的关节状态 -->
<node name = "joint_state_publisher" pkg = "joint_state_publisher" type = "joint_state_publisher" />

<!-- 运行 robot_state_publisher 节点,发布 tf -->
<node name = "robot_state_publisher" pkg = "robot_state_publisher" type = "robot_state_publisher" />

<!-- 运行 rviz 可视化界面 -->
<node name = "rviz" pkg = "rviz" type = "rviz" args = "-d $(find mbot_description)/config/mbot.rviz" required = "true" />

</launch>
```

运行以下命令打开 rviz(需要上面的 gazebo 也处于打开状态)：

```
$ roslaunch mbot_description display_mbot_xacro.launch
```

激光雷达数据如图 9.33 所示。

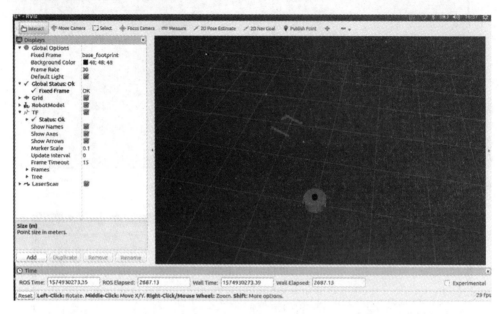

图 9.33　激光雷达数据

（3）添加 USB 相机

在 mbot_description/urdf/xacro/sensors 路径新建一个 xarco 文件命名为：camera_gazebo.xacro，并输入以下代码：

```xml
<?xml version = "1.0"?>
<robot xmlns:xacro = "http://www.ros.org/wiki/xacro" name = "camera">

    <xacro:macro name = "usb_camera" params = "prefix: = camera">
        <!-- 创建激光雷达参考系 -->
        <link name = "${prefix}_link">
            <inertial>
                <mass value = "0.1" />
                <origin xyz = "0 0 0" />
                <inertia ixx = "0.01" ixy = "0.0" ixz = "0.0"
                         iyy = "0.01" iyz = "0.0"
                         izz = "0.01" />
            </inertial>

            <visual>
                <origin xyz = " 0 0 0 " rpy = "0 0 0 " />
                <geometry>
                    <box size = "0.01 0.04 0.04" />
                </geometry>
                <material name = "black"/>
            </visual>
```

```xml
        <collision>
            <origin xyz = "0.0 0.0 0.0" rpy = "0 0 0" />
            <geometry>
                <box size = "0.01 0.04 0.04" />
            </geometry>
        </collision>
    </link>
    <gazebo reference = " $ {prefix}_link">
        <material> Gazebo/Black </material>
    </gazebo>

    <gazebo reference = " $ {prefix}_link">
        <sensor type = "camera" name = "camera_node">
            <update_rate> 30.0 </update_rate>
            <camera name = "head">
                <horizontal_fov> 1.3962634 </horizontal_fov>
                <image>
                    <width> 1280 </width>
                    <height> 720 </height>
                    <format> R8G8B8 </format>
                </image>
                <clip>
                    <near> 0.02 </near>
                    <far> 300 </far>
                </clip>
                <noise>
                    <type> gaussian </type>
                    <mean> 0.0 </mean>
                    <stddev> 0.007 </stddev>
                </noise>
            </camera>
            <plugin name = "gazebo_camera" filename = "libgazebo_ros_camera.so">
                <alwaysOn> true </alwaysOn>
                <updateRate> 0.0 </updateRate>
                <cameraName> /camera </cameraName>
                <imageTopicName> image_raw </imageTopicName>
                <cameraInfoTopicName> camera_info </cameraInfoTopicName>
                <frameName> camera_link </frameName>
                <hackBaseline> 0.07 </hackBaseline>
                <distortionK1> 0.0 </distortionK1>
                <distortionK2> 0.0 </distortionK2>
                <distortionK3> 0.0 </distortionK3>
                <distortionT1> 0.0 </distortionT1>
                <distortionT2> 0.0 </distortionT2>
            </plugin>
        </sensor>
    </gazebo>

    </xacro:macro>
</robot>
```

上述代码中包含了一个摄像头节点及一个摄像头插件。

在 gazebo 中显示添加 USB 摄像头后的机器人模型。修改 mbot_description/urdf/xacro/gazebo 下的 mbot_gazebo.xacro 文件如下：

```xml
<?xml version = "1.0"?>
<robot name = "arm" xmlns:xacro = "http://www.ros.org/wiki/xacro">

    <xacro:include filename = "$(find mbot_description)/urdf/xacro/gazebo/mbot_base_gazebo.xacro" />
    <xacro:include filename = "$(find mbot_description)/urdf/xacro/sensors/camera_gazebo.xacro" />

    <xacro:property name = "camera_offset_x" value = "0.17" />
    <xacro:property name = "camera_offset_y" value = "0" />
    <xacro:property name = "camera_offset_z" value = "0.10" />

    <mbot_base/>

    <!-- 摄像头 -->
    <joint name = "camera_joint" type = "fixed">
        <origin xyz = "${camera_offset_x} ${camera_offset_y} ${camera_offset_z}" rpy = "0 0 0" />
        <parent link = "base_link" />
        <child link = "camera_link" />
    </joint>

    <xacro:usb_camera prefix = "camera" />

    <mbot_base_gazebo/>

</robot>
```

运行以下命令打开 gazebo：

```
$ roslaunch mbot_description display_mbot_gazebo.launch
```

正常运行如图 9.34 所示。

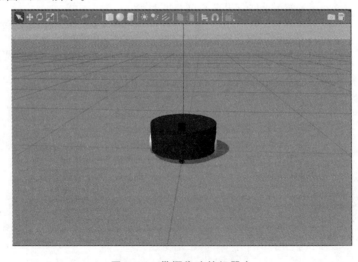

图 9.34　带摄像头的机器人

运行以下命令打开 rviz 查看摄像头的数据(需要上面的 gazebo 也处于打开状态):

$ roslaunch mbot_description display_mbot_xacro.launch

在 rviz 中添加 cameral 数据之后,如图 9.35 所示可以查看摄像头采集的数据。

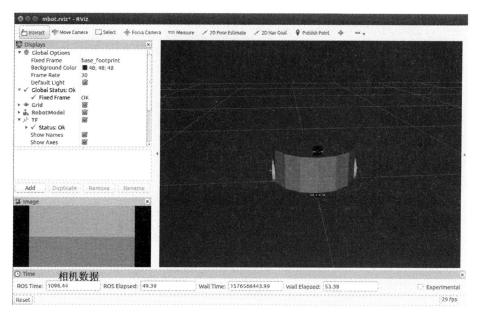

图 9.35　在 rviz 中查看摄像头数据

(4) 添加深度摄像头传感器

在 mbot_description/urdf/xacro/sensors 路径新建一个 xarco 文件命名为:kinect_gazebo.xacro,并输入以下代码:

```xml
<?xml version = "1.0"?>
<robot xmlns:xacro = "http://www.ros.org/wiki/xacro" name = "kinect_camera">

    <xacro:macro name = "kinect_camera" params = "prefix: = camera">
        <!-- 创建 kinect 参考系 -->
        <!-- 为 kinect 加入渲染 -->
        <link name = " ${prefix}_link">
            <origin xyz = "0 0 0" rpy = "0 0 0"/>
            <visual>
                <origin xyz = "0 0 0" rpy = "0 0 ${M_PI/2}"/>
                <geometry>
                    <mesh filename = "package://mbot_description/meshes/kinect.dae" />
                </geometry>
            </visual>
            <collision>
                <geometry>
                    <box size = "0.07 0.3 0.09"/>
                </geometry>
            </collision>
        </link>
```

```xml
<joint name="${prefix}_optical_joint" type="fixed">
    <origin xyz="0 0 0" rpy="-1.5708 0 -1.5708"/>
    <parent link="${prefix}_link"/>
    <child link="${prefix}_frame_optical"/>
</joint>

<link name="${prefix}_frame_optical"/>
<!-- 以下为深度摄像头插件 -->
<gazebo reference="${prefix}_link">
    <sensor type="depth" name="${prefix}">
        <always_on> true </always_on>
        <update_rate> 20.0 </update_rate>
        <camera>
            <horizontal_fov> ${60.0 * M_PI/180.0} </horizontal_fov>
            <image>
                <format> R8G8B8 </format>
                <width> 640 </width>
                <height> 480 </height>
            </image>
            <clip>
                <near> 0.05 </near>
                <far> 8.0 </far>
            </clip>
        </camera>
        <plugin name="kinect_${prefix}_controller" filename="libgazebo_ros_openni_kinect.so">
            <cameraName> ${prefix} </cameraName>
            <alwaysOn> true </alwaysOn>
            <updateRate> 10 </updateRate>
            <imageTopicName> rgb/image_raw </imageTopicName>
            <depthImageTopicName> depth/image_raw </depthImageTopicName>
            <pointCloudTopicName> depth/points </pointCloudTopicName>
            <cameraInfoTopicName> rgb/camera_info </cameraInfoTopicName>
            <depthImageCameraInfoTopicName> depth/camera_info </depthImageCameraInfoTopicName>
            <frameName> ${prefix}_frame_optical </frameName>
            <baseline> 0.1 </baseline>
            <distortion_k1> 0.0 </distortion_k1>
            <distortion_k2> 0.0 </distortion_k2>
            <distortion_k3> 0.0 </distortion_k3>
            <distortion_t1> 0.0 </distortion_t1>
            <distortion_t2> 0.0 </distortion_t2>
            <pointCloudCutoff> 0.4 </pointCloudCutoff>
        </plugin>
    </sensor>
</gazebo>

</xacro:macro>
</robot>
```

上述代码中包含了深度摄像头的节点及插件。

在 gazebo 中显示添加 USB 摄像头后的机器人模型。修改 mbot_description/urdf/xacro/

gazebo 下的 mbot_gazebo.xacro 文件如下：

```xml
<?xml version = "1.0"?>
<robot name = "arm" xmlns:xacro = "http://www.ros.org/wiki/xacro">

    <xacro:include filename = "$(find mbot_description)/urdf/xacro/gazebo/mbot_base_gazebo.xacro" />
    <xacro:include filename = "$(find mbot_description)/urdf/xacro/sensors/kinect_gazebo.xacro" />

    <xacro:property name = "kinect_offset_x" value = "0.15" />
    <xacro:property name = "kinect_offset_y" value = "0" />
    <xacro:property name = "kinect_offset_z" value = "0.11" />

    <mbot_base/>

    <!-- kinect -->
    <joint name = "kinect_joint" type = "fixed">
        <origin xyz = "${kinect_offset_x} ${kinect_offset_y} ${kinect_offset_z}" rpy = "0 0 0" />
        <parent link = "base_link"/>
        <child link = "kinect_link"/>
    </joint>

    <xacro:kinect_camera prefix = "kinect"/>

    <mbot_base_gazebo/>

</robot>
```

运行以下命令查看 gazebo 中的机器人模型（如图 9.36 所示）：

$ roslaunch mbot_description display_mbot_gazebo.launch

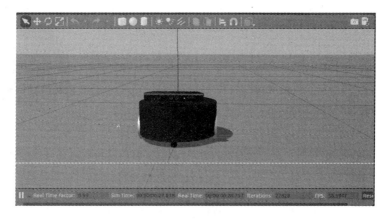

图 9.36　带有 kinect 深度摄像头的机器人模型

打开一个终端输入以下命令查看深度摄像头发布的话题，如图 9.37 所示。
当然，我们也可以在 rviz 中查看 Kinect 的可视化数据，方法与激光雷达及相机相同。

```
/clock
/cmd_vel
/gazebo/link_states
/gazebo/model_states
/gazebo/parameter_descriptions
/gazebo/parameter_updates
/gazebo/set_link_state
/gazebo/set_model_state
/gazebo_gui/parameter_descriptions
/gazebo_gui/parameter_updates
/joint_states
/kinect/depth/camera_info
/kinect/depth/image_raw
/kinect/depth/points
/kinect/parameter_descriptions
/kinect/parameter_updates
/kinect/rgb/camera_info
/kinect/rgb/image_raw
/kinect/rgb/image_raw/compressed
/kinect/rgb/image_raw/compressed/parameter_descriptions
/kinect/rgb/image_raw/compressed/parameter_updates
/kinect/rgb/image_raw/compressedDepth
/kinect/rgb/image_raw/compressedDepth/parameter_descriptions
/kinect/rgb/image_raw/compressedDepth/parameter_updates
/kinect/rgb/image_raw/theora
/kinect/rgb/image_raw/theora/parameter_descriptions
/kinect/rgb/image_raw/theora/parameter_updates
/odom
/rosout
/rosout_agg
/tf
/tf_static
```

图 9.37 深度摄像头发布的话题

9.7 sw2urdf 工具

在前面学习了如何利用 URDF 描述语言建立一个机器人模型，如何建立 gazebo 模型，如何添加控制器及传感器。现实工作中，我们很难用简单的机构形状来描述机器人，如图 9.38 所示的 bobac 智能服务机器人，很难用常规的 URDF 语言描述。下面介绍一个工具 sw2urdf。sw2urdf 是一个 solidworks 的插件，可以将 solidworks 的三维模型直接转化为 URDF 文件。

下载 sw2urdf 插件：http://wiki.ros.org/sw_urdf_exporter。

使用 sw2urdf：http://wiki.ros.org/sw_urdf_exporter/Tutorials。

图 9.38　bobac 智能机器人

在本书的随书资料中有一个利用 sw2urdf 插件生成的 URDF 机器人描述包：bobac_description。将 bobac_description 复制到 ros_workspace/src 目录下编译。

运行以下命令在 gazebo 中显示 bobac 智能服务机器人（如图 9.39 所示）：

$ roslaunch bobac2_description gazebo.launch

bobac 智能服务机器人的仿真模型中包含了一个两轮差速控制器、一个二维激光雷达、一个深度摄像头。

图 9.39 gazebo 中的 bobac 智能服务机器人

课后练习

一、选择题

(1) [单选] gazebo 是一款(　　)工具？

　　(A) 调试

　　(B) 可视化

　　(C) 仿真

　　(D) 命令行

(2) [单选] rqt_graph 可以用来查看计算图，以下说法错误的是(　　)。

　　(A) 计算图反映了节点之间消息的流向

　　(B) rqt_graph 中的椭圆代表节点

　　(C) rqt_graph 可以看到所有的话题、服务和动作通信方式

　　(D) 计算图反映了所有运行的节点

(3) [单选] ROS 中 Odom 里程计的数据来源包括(　　)。

　　(A) Visual Odometry

　　(B) IMU

　　(C) Wheel Odometry

　　(D) Laser

(4) [多选] rviz 可以图形化显示的类型的数据(　　)。

　　(A) 激光 LaserScan

　　(B) 点云 PointCloud

　　(C) 机器人模型 RobotModel

　　(D) 轨迹 Path

二、操作题

创建一个简易的仿真机器人,通过使用 gazebo 来获得与现实相同的传感器数据,利用激光雷达搜索并捕捉目标,编写 ROS 节点实现以下功能:

(1)当机器人激光雷达在规定范围内扫描到物体时,程序通过订阅仿真出来的激光话题,获取距离最近的障碍物的位置,然后发布相应的速度话题;

(2)仿真机器人便会自动订阅该速度话题,最终到达该障碍物附近。

第 10 章 常见运动学解算

运动学涉及机器人机构中物体的运动,但并不考虑引起运动的力/力矩。由于机器人机构是为运动而精心设计的,所以运动学是机器人的设计、分析、控制和仿真的基础。

在本章中讨论几种常见的轮式移动机器人的运动学分析。移动机器人的运动学分析分为运动学正向解和运动学反向解。

运动学正向解是一个"观测"问题,通过编码器这种简单的传感器可以方便地测出每个轮子的转速$(n_1, n_2, \cdots, n_n)$,然后通过运动学正向解得到机器人运动的线速度v及角速度ω,如图 10.1 所示。

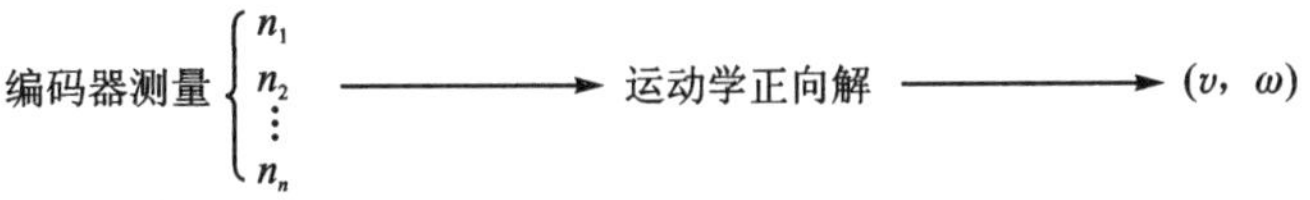

编码器测量 $\begin{cases} n_1 \\ n_2 \\ \vdots \\ n_n \end{cases}$ ⟶ 运动学正向解 ⟶ (v, ω)

图 10.1 运动学正向解

运动学反向解是一个"控制"问题,对于车体移动的控制,我们期望控制量是车体运动的线速度(v)和角速度(ω),但是实际中能够直接控制的是每个车轮电机的转速$(n_1, n_2, \cdots, n_n)$。运动学反向解正是连接(v, ω)到$(n_1, n_2, \cdots, n_n)$的纽带,如图 10.2 所示。

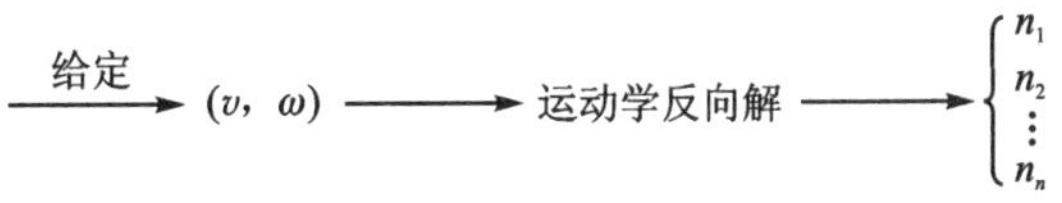

给定 ⟶ (v, ω) ⟶ 运动学反向解 ⟶ $\begin{cases} n_1 \\ n_2 \\ \vdots \\ n_n \end{cases}$

图 10.2 运动学反向解

10.1 两轮差动模型

两轮差动模型是最常见的移动机器人运动模型之一,两轮差动机器人由两个驱动轮及一个或多个万向轮组成。如图 10.3 所示为两种比较典型的两轮差动模型。

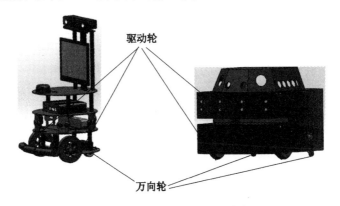

图 10.3 常见的两种两轮差动模型

1. 两轮差动运动学正向解

如图 10.4 所示是机器人在相邻两个时刻的位姿。

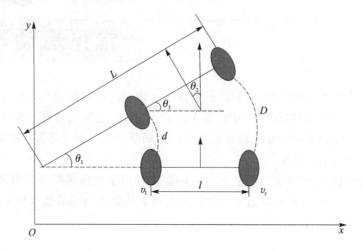

图 10.4 机器人在相邻两个时刻的位姿

图 10.4 中：

l：两轮之间的间距；

L：在 Δt 期间机器人的旋转半径；

v_l：左轮线速度；

v_r：右轮线速度；

θ_2：机器人旋转角度；

d：左轮在 Δt 期间的位移；

D：右轮在 Δt 期间的位移。

显然，机器人移动的线速度 v 为左右轮线速度的平均值

$$v = \frac{v_l + v_r}{2} \tag{10.1}$$

由图 10.4 中的几何关系可以得到下式：

$$\theta_1 = \theta_2 = \theta_3 \tag{10.2}$$

左右两轮在 Δt 期间的位移差为

$$D - d = (v_r - v_l) \cdot \Delta t \tag{10.3}$$

其中：

$$D = L \cdot \theta_1 \tag{10.4}$$

$$d = (L - l) \cdot \theta_1 \tag{10.5}$$

由式(10.3)~式(10.5)可得：

$$(L - l) \cdot \theta_1 + (v_r - v_l) \cdot \Delta t = L \cdot \theta_1 \tag{10.6}$$

由式(10.6)可得：

$$\frac{\theta_1}{\Delta t} = \frac{v_r - v_l}{l} = \omega \tag{10.7}$$

故两轮差动机器人的线速度和角速度，即两轮差动模型的运动学正向解为

$$\left.\begin{aligned}\omega &= \frac{v_\mathrm{r}-v_\mathrm{l}}{l}\\ v &= \frac{v_\mathrm{l}+v_\mathrm{r}}{2}\end{aligned}\right\} \tag{10.8}$$

2. 两轮差动运动学反向解

将式(10.8)中的运动学正向解进行反运算,可得反向解为

$$\left.\begin{aligned}v_\mathrm{r} &= v + \frac{\omega l}{2}\\ v_\mathrm{l} &= v - \frac{\omega l}{2}\end{aligned}\right\} \tag{10.9}$$

10.2 三轮全向运动模型

三轮全向移动底盘因其良好的运动性且结构简单,近年来备受欢迎。三个轮子互相间隔120°,每个全向轮由若干个小滚轮组成,各个滚轮的母线组成一个完整的圆。机器人既可以沿轮面的切线方向移动,也可以沿轮子的轴线方向移动,这两种运动的组合即可实现平面内任意方向的运动。

如图10.5所示为典型的三轮全向机构及一个以三轮全向为底盘的机器人。

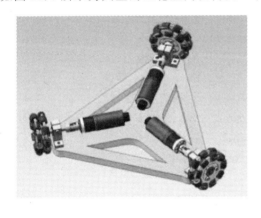

图10.5 三轮全向底盘及三轮全向机器人

1. 运动学反向解

为便于运动学分析,我们以理想情况为基础,三个轮子相对于车体的中轴线对称,且物理尺寸、质量等完全一致;上层负载均衡,机器人的重心与三个轮子转动轴线的交点重合;三个轮体与地面摩擦力足够大,不会发生打滑现象;机器人中心到三个全向轮的距离相等。

三轮全向底盘运动学分析如图10.6所示。

图中:

xOy 机器人自身坐标系;

v_y 机器人沿自身坐标系 y 方向移动的速度;

v_x 机器人沿自身坐标系 x 方向移动的速度;

v_θ 机器人绕自身中心旋转速度;

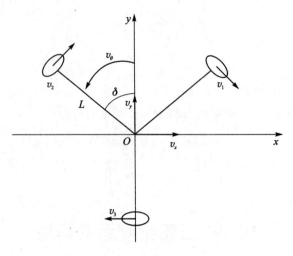

图 10.6 三轮全向底盘运动学分析

L 轮子到底盘中心之间的距离；

v_1, v_2, v_3 三个轮子的线速度；

δ 轮子 3 与 y 轴正方向夹角，这里 $\delta = 60°$。

约定逆时针旋转为正。将轮 1 的线速度 v_1 分解到机器人自身坐标的 x, y 轴上可得：

$$v_1 = v_x \cdot \cos\delta - v_y \cdot \sin\delta - L \cdot v_\theta \tag{10.10}$$

同理可得：

$$v_2 = v_x \cdot \cos\delta + v_y \cdot \sin\delta - L \cdot v_\theta \tag{10.11}$$

$$v_3 = -v_x - L \cdot v_\theta \tag{10.12}$$

写成矩阵的形式为

$$\begin{bmatrix} v_1 \\ v_2 \\ v_3 \end{bmatrix} = \begin{bmatrix} \cos\delta & -\sin\delta & -L \\ \cos\delta & \sin\delta & -L \\ -1 & 0 & -L \end{bmatrix} \begin{bmatrix} v_x \\ v_y \\ v_\theta \end{bmatrix} \tag{10.13}$$

将 $\delta = 60°$ 代入得：

$$\begin{bmatrix} v_1 \\ v_2 \\ v_3 \end{bmatrix} = \begin{bmatrix} \dfrac{1}{2} & -\dfrac{\sqrt{3}}{2} & -L \\ \dfrac{1}{2} & \dfrac{\sqrt{3}}{2} & -L \\ -1 & 0 & -L \end{bmatrix} \begin{bmatrix} v_x \\ v_y \\ v_\theta \end{bmatrix} \tag{10.14}$$

2. 运动学正向解

将式（10.14）求逆运算得到三轮全向的运动学正向解为

$$\begin{bmatrix} v_x \\ v_y \\ v_\theta \end{bmatrix} = \begin{bmatrix} \dfrac{1}{3} & \dfrac{1}{3} & -\dfrac{2}{3} \\ -\dfrac{1}{\sqrt{3}} & \dfrac{1}{\sqrt{3}} & 0 \\ -\dfrac{1}{3L} & -\dfrac{1}{3L} & -\dfrac{1}{3L} \end{bmatrix} \begin{bmatrix} v_1 \\ v_2 \\ v_3 \end{bmatrix} \tag{10.15}$$

10.3 四轮全驱滑动运动模型

四轮全驱滑移模型是一种比较常见的户外运动车体模型,具有动力足、转弯半径小等特点。图10.7所示为四轮全驱滑移运动模型车体。

图10.7 四轮全驱滑移运动模型实物图

1. 四轮全驱滑移运动学正向解

四轮全驱滑移运动学分析与两轮差动运动学分析相似,如图10.8所示为相邻两个时刻机器人的位姿,两个时刻相差的时间间隔为 Δt。

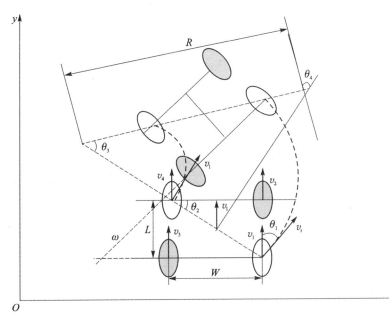

图10.8 四轮全驱滑移运动模型分析图

图中：

xOy　车体移动的全局坐标系；

v_1, v_2, v_3, v_4　机器人四个轮子的线速度；

v　机器人移动的线速度；

v_r　机器人右侧绕中心旋转的线速度；

v_l　机器人左侧绕中心旋转的线速度；

ω　机器人旋转的角速度；

L　机器人前后轮之间的间距；

W　机器人左右轮之间的间距；

R　机器人在 Δt 期间的旋转半径；

θ_1　机器人轮子线速度与旋转线速度之间的夹角；

θ_2　前后轮对角线与两个前轮之间的夹角；

θ_3　在 Δt 期间机器人旋转的角度；

θ_4　相邻时刻两轮子航向角的增量。

为了达到最优控制效果，四轮滑移模型的控制方法采用左边两轮的速度相等，右边两个轮子的速度相等，通过左右轮的差速实现车体的转动，即

$$\left.\begin{aligned} v_1 &= v_2 \\ v_3 &= v_4 \end{aligned}\right\} \quad (10.16)$$

显然机器人线速度 v 是左右轮速度的平均值，即

$$v = \frac{v_1 + v_4}{2} \quad (10.17)$$

车体右侧绕车体中心旋转的线速度表达式为

$$v_r = v_1 \cdot \cos\theta_1 \quad (10.18)$$

由于 $\theta_1 = \theta_2$，故

$$\cos\theta_1 = \frac{W}{\sqrt{W^2 + L^2}} \quad (10.19)$$

所以

$$v_r = v_1 \cdot \frac{W}{\sqrt{W^2 + L^2}} \quad (10.20)$$

同理，车体左侧绕车体中心旋转的线速度 v_l 的表达式为

$$v_l = v_4 \cdot \frac{W}{\sqrt{W^2 + L^2}} \quad (10.21)$$

同理，两轮差动模型的运动学正向解运算得到车体旋转的角速度为

$$\omega = \frac{v_r - v_l}{\sqrt{W^2 + L^2}} \quad (10.22)$$

即

$$\omega = \frac{W \cdot (v_1 - v_4)}{W^2 + L^2} \quad (10.23)$$

综上可得，四轮全驱滑移模型的运动学正向解为

$$\left.\begin{array}{l}\omega = \dfrac{W \cdot (v_1 - v_4)}{W^2 + L^2} \\ v = \dfrac{v_1 + v_4}{2}\end{array}\right\} \quad (10.24)$$

2. 四轮全驱滑移运动学反向解

将式(10.24)做运动学逆运算,可得四轮全驱滑移模型的运动学反向解为

$$\left.\begin{array}{l}v_1 = v + \dfrac{W^2 + L^2}{2W}\omega \\ v_4 = v - \dfrac{W^2 + L^2}{2W}\omega \\ v_2 = v_1 \\ v_3 = v_4\end{array}\right\} \quad (10.25)$$

10.4 四轮全向运动模型

四轮全向机器人一般采用麦克纳姆轮作为驱动轮,动力方面采用四轮全驱的方式。麦克纳姆轮由两大部分组成:轮毂和辊子(roller)。轮毂是整个轮子的主体支架,辊子则是安装在轮毂上的鼓状物。麦克纳姆轮的轮毂轴与辊子转轴成 45°角。理论上,这个夹角可以是任意值,根据不同的夹角可以制作出不同的轮子,但最常用的还是 45°。如图 10.9 所示为一个典型的麦克纳姆轮(简称麦轮)。

图 10.9 典型麦克纳姆轮

麦轮一般是 4 个一组使用,两个左旋轮,两个右旋轮。左旋轮和右旋轮呈手性对称,如图 10.10 所示。

图 10.10 麦克纳姆轮左右旋轮

安装方式有多种,主要分为:X-正方形(X-square)、X-长方形(X-rectangle)、O-正方形(O-square)、O-长方形(O-rectangle)。其中:X 和 O 表示 4 个轮子与地面接触的辊子所形成的图形;正方形与长方形指 4 个轮子与地面接触点所围成的形状。如图 10.11 所示分别为:X-正方形、X-长方形、O-正方形、O-长方形。

X-正方形:轮子转动产生的力矩会经过同一个点,所以 yaw 轴无法主动旋转,也无法主

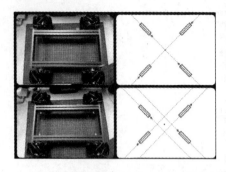
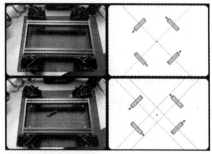

图 10.11 麦克纳姆轮 4 种安装方式

动保持 yaw 轴的角度。一般不会使用这种安装方式。

X-长方形:轮子转动可以产生 yaw 轴转动力矩,但转动力矩的力臂一般会比较短。这种安装方式也不多见。

O-正方形:4 个轮子位于正方形的 4 个顶点,平移和旋转都没有任何问题。受限于机器人底盘的形状、尺寸等因素,这种安装方式虽然理想,但可遇而不可求。

O-长方形:轮子转动可以产生 yaw 轴转动力矩,而且转动力矩的力臂也比较长,是最常见的安装方式。

下面介绍最常见的 O-长方形安装方式的运动学正反向解。

1. 四轮全向底盘运动学反向解

由于麦轮底盘的数学模型比较复杂,我们在此分 4 步进行:

① 将底盘的运动分解为 3 个独立变量来描述;
② 根据第①步的结果,计算出每个轮子轴心位置的速度;
③ 根据第②步的结果,计算出每个轮子与地面接触的辊子的速度;
④ 根据第③部的结果,计算出轮子的真实转速。

(1) 底盘运动分解

我们知道,刚体在平面内的运动可以分解为 3 个独立分量:x 轴平动、y 轴平动、yaw 轴自转。如图 10.12 所示,底盘的运动也可以分解为 3 个量:

v_{tx} 表示 x 轴运动的速度,即左右方向,定义向右为正;
v_{ty} 表示 y 轴运动的速度,即前后方向,定义向前为正;
ω 表示 yaw 轴自转的角速度,定义逆时针为正。

以上 3 个量一般都视为四个轮子的几何中心(矩形的对角线交点)的速度。

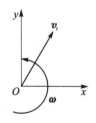

图 10.12 底盘在平面上运动分解

(2) 计算出轮子轴心位置的速度

定义:

r 为从几何中心指向轮子轴心的矢量;
v 为轮子轴心的运动速度矢量;
v_t 为轮子轴心沿垂直于 r 的方向(即切线方向)的速度分量。

轮 1 轴心位置的速度计算如图 10.13 所示。

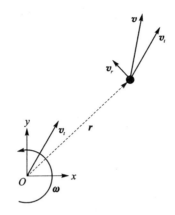

图 10.13 轮 1 轴心位置的速度计算

那么可以计算出：

$$v = v_t + \omega \times r \quad (10.26)$$

分别计算 x、y 轴的分量为

$$\begin{cases} v_x = v_{tx} - \omega \cdot r_y \\ v_y = v_{ty} + \omega \cdot r_x \end{cases} \quad (10.27)$$

同理，可以算出其他 3 个轮子轴心的速度，如图 10.14 所示。

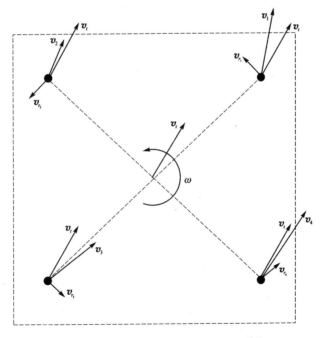

图 10.14 4 个轮子轴心位置的速度计算

(3) 计算辊子的速度

根据轮子轴心的速度，可以分解出沿辊子方向的速度 $v_\parallel$ 和垂直于辊子方向的速度 $v_\perp$，如图 10.15 所示。其中，$v_\perp$ 是可以忽视的（很简单，自己思考一下哈），而

$$v_{\parallel} = \boldsymbol{v} \cdot \hat{\boldsymbol{u}} = (v_x \hat{\boldsymbol{i}} + v_y \hat{\boldsymbol{j}}) \cdot \left(-\frac{1}{\sqrt{2}} \hat{\boldsymbol{i}} + \frac{1}{\sqrt{2}} \hat{\boldsymbol{j}} \right) = -\frac{1}{\sqrt{2}} v_x + \frac{1}{\sqrt{2}} v_y \qquad (10.28)$$

式中,$\hat{\boldsymbol{u}}$ 是沿辊子方向的单位矢量。

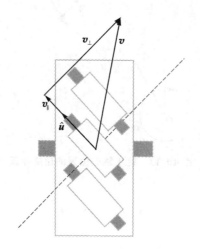

图 10.15　计算辊子的速度

(4) 计算轮子速度

从辊子速度到轮子转速的计算比较简单,如图 10.16 所示。其表达为

$$v_\omega = \frac{v_{\parallel}}{\cos 45°} = \sqrt{2} \left(-\frac{1}{\sqrt{2}} v_x + \frac{1}{\sqrt{2}} v_y \right) = -v_x + v_y \qquad (10.29)$$

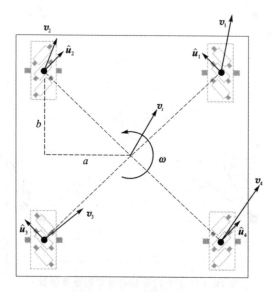

图 10.16　计算轮子速度

根据图 10.16 中 a,b 的定义以及式(10.8),有：

$$\begin{cases} v_x = v_{tx} - \omega \cdot b \\ v_y = v_{ty} + \omega \cdot a \end{cases} \qquad (10.30)$$

将式(10.30)代入式(10.29)得到轮1的线速度为

$$v_\omega = v_{ty} - v_{tx} + \omega(a+b) \tag{10.31}$$

同理,可以得到其他3个轮的线速度(过程大家可以自己推理一下)。四轮全向运动模型的运动学反向解为

$$\begin{cases} v_{\omega 1} = v_{ty} - v_{tx} + \omega(a+b) \\ v_{\omega 2} = v_{ty} + v_{tx} - \omega(a+b) \\ v_{\omega 3} = v_{ty} - v_{tx} - \omega(a+b) \\ v_{\omega 4} = v_{ty} + v_{tx} + \omega(a+b) \end{cases} \tag{10.32}$$

2. 四轮全向底盘运动学正向解

将式(10.32)中的3个方程求逆运算,可得到四轮全向底盘运动学正向解为

$$\begin{cases} v_{tx} = \dfrac{v_{\omega 4} - v_{\omega 1}}{2} \\ v_{ty} = \dfrac{v_{\omega 2} + v_{\omega 1}}{2} \\ \omega = \dfrac{v_{\omega 4} - v_{\omega 2}}{a+b} \end{cases} \tag{10.33}$$

在本章中对常见的4种运动模型进行了运动学分析,在运动学分析的基础上编写程序控制机器人运动以及获取机器人运动的里程信息就容易得多了。

课后练习

一、选择题

[多选] 如图10.17所示差速底盘有以下4种类型,其中图(a)、(b)的底盘轮廓是圆形,而图(c)、(d)的底盘轮廓是矩形。不同的构形在机器人运动稳定性、负载能力等方面有着不同的表现,其应用场景也有区别,下列叙述正确的是(　　)。

(A) 图(a)中的构形是前后采用万向轮,两侧是驱动轮,4个轮子需要使用悬挂,否则容易出现任一轮子悬空打滑的情况

(B) 图(b)相比图(a)中去掉了1个万向轮,也就无需使用悬挂,为了弥补稳定性,需将两驱动轮后移一小段距离

(C) 图(c)将两个万向轮前置,驱动轮后置,车上的长度可以较长,但也需要悬挂系统。由于旋转中心和重心相差较远,因此无人车旋转运动性能偏弱

(D) 图(d)使用了4个万向轮,布置于前后,在对称轴处布置驱动轮,这样便保证旋转中心和重心重合,且能做到车身更长

二、简答题

请叙述两轮差运模型的推导过程?

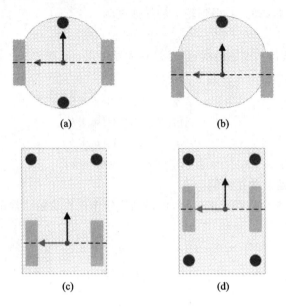

注:黑圆点表示万向轮,矩形块表示驱动轮。

图 10.17　不同类型的两轮差速平台

参考文献

[1] 李克强,戴一凡,李升波,等. 智能网联汽车(ICV)技术的发展现状及趋势[J]. 汽车安全与节能学报,2017(1):1-14.

[2] ROS wiki[EB/OL]. [2021-5-21] http://wiki.ros.org.

[3] 崔胜民. 智能网联汽车新技术[M]. 北京:化学工业出版社,2016.

[4] 徐可,徐楠. 全球视角下的智能网联汽车发展路径[J]. 中国工业评论,2015(9):76-82.

[5] 陈慧,涂强. 互联智能汽车关键技术与发展趋势[J]. 中国集成电路,2015(6):24-30.

[6] 张亚萍,刘华,等. 智能网联汽车技术与标准发展研究[J]. 上海汽车,2015(8):55-59.

[7] 谢伯元,李克强,等. "三网融合"的车联网概念及其在汽车工业中的应用[J]. 汽车安全与节能学报,2013,4(4):348-355.

[8] 陈慧岩,熊光明,等. 无人驾驶汽车概论[M]. 北京:北京理工大学出版社,2014.

[9] Azim Eskandarian,李克强. 智能车辆手册[M]. 北京:机械工业出版社,2017.

[10] Carol Fairchild,Thomas L. Harman. ROS Robotics By Example[M]. Birmingham:Packt Publishing,2017.

[11] AndreaTesta,Andrea Camisa,Giuseppe Notarstefano. ChoiRbot:A ROS 2 Toolbox for Cooperative Robotics[EB/OL]. https://arxiv.org/pdf/2010.13431.pdf.

[12] 胡春旭. ROS机器人开发实践[M]. 北京:机械工业出版社,2018.

[13] Anil Mahtani,Luis Sanchez,Enrique Fernandez,et al. Effective Robotics Programming with ROS[M]. Birmingham:Packt Publishing,2018.

[14] 杨成. 无人驾驶智能车障碍检测方法研究[D]. 西安:西安工业大学,2015.

[15] 刘丹. 智能车辆同时定位与建图关键技术研究[D]. 北京:北京工业大学,2018.

[16] 王俊. 无人驾驶车辆环境感知系统关键技术研究[D]. 合肥:中国科学技术大学,2016.

[17] 刘伟. 基于激光雷达和机器视觉的智能车前方障碍物检测研究[D]. 哈尔滨:哈尔滨理工大学,2019.

[18] 方海洋. 基于GPS和激光雷达的无人驾驶策略研究[D]. 西安:长安大学,2019.

[19] 叶伟铨. 无人车的自主导航与控制研究[D]. 广州:华南理工大学,2016.

[20] 王建强,王昕. 智能网联汽车体系结构与关键技术[J]. 长安大学学报(社会科学版),2017(6):18-25.